T0161013

Rough Likeness

Rough Likeness

essays

Lia Purpura

Sarabande Books
LOUISVILLE, KENTUCKY

FIRST EDITION

Managing Editor
Sarabande Books, Inc.
2234 Dundee Road, Suite 200
Louisville, KY 40205

Library of Congress Cataloging-in-Publication Data

Purpura, Lia.
Rough likeness : essays / Lia Purpura.
p. cm.
ISBN 978-1-936747-03-0 (pbk. : alk. paper)
I. Title.
PS3566.U67R68 2011
814'.54—dc22
2011007028

Cover and text design by Kirkby Gann Tittle.

Manufactured in Canada.
This book is printed on acid-free paper.

Sarabande Books is a nonprofit literary organization.

This project is supported in part by an award from the National Endowment for the Arts.

The Kentucky Arts Council, the state arts agency, supports Sarabande Books with state tax dollars and federal funding from the National Endowment for the Arts.

For Jed and Joseph

Also by Lia Purpura

King Baby (poems)
On Looking (essays)
Stone Sky Lifting (poems)
Increase (essays)
Poems of Grzegorz Musial (translations)
The Brighter the Veil (poems)

CONTENTS

ACKNOWLEDGMENTS

I'd like to thank the editors of the following magazines in which these essays first appeared, sometimes in slightly different form:

Agni: "Against 'Gunmetal'"; "Being of Two Minds"; "The Lustres"; "Memo Re: Beach Glass"; "Two Experiments & a Coda"
Arts & Letters: "'Try Our Delicious Pizza'"
Black Warrior Review: "'Poetry Is a Satisfying of the Desire for Resemblance' (Theme & Variations)"
Crazyhorse: "Street Scene"
Ecotone: "On Tools"
Iowa Review: "Augury"; "Jump"
Ninth Letter: "Remembering"
Orion: "On Coming Back as a Buzzard"; "There Are Things Awry Here"
Seattle Review: "Advice"
Sonora Review: "On Luxury"
Southern Review: "Gray"

"There Are Things Awry Here" was reprinted in *Best American Essays, 2011;* "On Coming Back as a Buzzard" and "Two Experiments & a Coda" were reprinted in *The Pushcart Anthology XXXV 2011* and *XXXIV 2010.* "The Lustres" and "Remembering" were named "Notable Essays" in *Best American Essays 2008* and *2009.*

I am deeply grateful to Jed Gaylin, Maddalena Purpura, and Kent Meyers for advice of the revelatory variety, and for the lavish attention they've given to my work; to Dan Corrigan for fine tracking skills; Alan Kolc, for artistry; Bill Pierce and Hilarie Gaylin for pinch hitting; and to Loyola University's Center for Humanities for generous summer study grants. And to Sarah Gorham, whose keen eye, light touch, and sustaining faith are rare and wondrous gifts. Daily, I know how lucky I am to have found her.

Rough Likeness

On Coming Back as a Buzzard

I know, coming back as a crow is a lot more attractive. If crows and buzzards do the same rough job—picking, tearing and cleaning up—who wouldn't rather return as a shiny blue crow with a mind for locks and puzzles. A strong voice, and poem-struck. Sleek, familial, omen-bearing. Full of mourning and ardor and talk. Buzzards are nothing like this, but something other, complicated by strangeness and ugliness. They intensify my thinking. They look prehistoric, pieced together, concerned. I might simply say I feel closer to them—always have—and proceed. Because, really, as I turn it over, the problem I'm working on here, coming back as a buzzard, has not so much to do with buzzards after all.

A buzzard is *expected* at the table. The rush would be over by the time I got there and I, my lateness sanctioned, might rightfully slip in. I wouldn't saunter, nor would I blow in dramatically—*flounce*, as my grandmother would say. The road would be the dinner table (just as the dinner table with its veering discussions, is always a road somewhere) and others' distraction would resolve—well, I would resolve it—into a clean plate.

I would be missed if I were not there. Not at first, not in the frenzy, but later.

Without me, no outlines, no profiles come clear. The very idea of scaffolding is diminished.

"The smaller scraps are tastier" would have no defender. "Close to the bone" would fall out of use as a measure of sharply felt truth.

Without a chance to walk away from abundance, thus proving their wealth, none of the first eaters would be content with their portion. I make their bestowing upon the least of us possible.

With me around, mishaps—side of the highway, over a cliff, more slowly dispensed by poison—do not have to be turned to a higher purpose. I step in. I make use of.

And here, I'm whittling away at the problem.

As a buzzard, I'd know the end of a thing is precisely not that. Things go on, in their way. My presence making the end a beginning, reinterpreting the idea of abundance, allowing for the ever-giving nature of Nature—I'd know these not as religious thoughts. It's that, apportioned rightly, there's always enough, more than enough. "Nothing but gifts on this poor, poor earth" says Milosz, who understood perfectly the resemblance between dissolve and increase. Rain scours and sun burns away excesses of form. And rain also seeds, and sun urges forth fuses of green.

I'd love best the movement of stages and increments, to repeat "this bank and shoal of time" while below me banks and shoals of a body went on welling/receding, rising and dropping. I'd be perched on a wire, waiting, ticking off not the meat reducing, but how what's left, like a dune, shifts and reconstitutes. Yes, it *looks* like I hover, and the hovering, I know, suggests a discomfiting eagerness. Malevolence. Why is that? I haven't killed a thing. If the waiting seems untoward, it may be confirming something too real, too true: all the parts that slip from sight, can't be easily had, collapse in on themselves and require digging, all the parts that promise small, intense bursts of sweetness unnerve us—while the easily abundant, the spans, the expanses (thick haunch, round belly and shoulder), all that lifts easily to another's lips and retains its form till the end—seems pure. Right and deserved. Proper and lawful. Thus butchers have their neat diagrams. One knows to call for *chop, loin, shank, rump.*

I'd get to be one, who, when passed the plate, seeks first the succulent eye. This would mark me: *foreigner*. Stubborn lover of scraps and dark meat. Base. Trained on want and come to love piecemeal offerings—the shreds and overlooked tendernesses too small for a meal, but carefully, singularly gathered—like brief moments that burst: isolate beams of sun in truck fumes, underside of wrist against wrist, sudden cool from a sewer grate rising. I incline toward the tucked and folded parts—it's that the old country can't be bred out—the internals with names that lack correspondence, the sweetbreads and umbles, bungs, hoods, and liver-and-lights. If the road is a plate, then the outskirts of fields and settlements where piles are heaped are plates, too. And the gullies, the ditches, the alleys—all plates. I'd get to reorder your thoughts about troves, to prove the spilled and shoveled-aside to be treasure. To reconfer notions of milk and honey, and how to approach the unbidden.

I resemble, as I suppose we all do, the things I consume: bent to those raw flaps of meat, red, torn, cast aside, my head also looks like a leftover thing, chewed. I have my ways of avoiding attention: vomit to turn away predators. Shit, like the elegant stork, on my legs to cool off, to disinfect the swarming microbes I tread daily. I am gentle. And cautious. I ride the thermals and flap very little (conserve, conserve) and locate food by smell. I'm a black V in air. A group of us on the ground is a *venue*. In the air we're a *kettle*.

I reuse even the language.

A simple word, *aftermath*, structures my day. Sometimes I think *epic*—doesn't everyone apply to their journey a story? Then *flyblown, feculent, scavenge* come—how it must seem to others—and the frame of my story's reduced. Things are made daily again. The first eaters are furiously driven—by hunger, and brute need releasing trap doors in the brain. Such push and ambition! I hold things in pantry spots in my body and take out and eat what I've saved when I need it and so am never furious. On my plate, choice reduces. I take what I come upon, and the work of a breeze cools the bowl's steaming contents. There's a beauty in this singularity: consider bringing to each occasion your one perfect bowl, one neat fork/spoon/knife set. That when the chance comes, you're given to draw the tine-curves between lips, pull, lick, tap clean the spoon's curvature—and for these sensations, there's ample time. Time pinned open, like the core of long summer afternoon.

Am I happy? Yes, in momentary ways. Which I think is a good way to feel about things that come when they will, and not when you will them. While I'm waiting, I get to be with the light as it shifts off the wet phone wire, catches low sun, holds, pearls and unpearls drops of water. If I bounce just a little, they shiver and fall, and my weight calls more pearls to me. There's light over the blood-matted rib-fur, and higher up, translucing on the still-unripped ear of the fox. Light through drops of fresh resin on pine limbs, light on ditchwater neverminding the murk. I get fixed by spoors of light, silver shine on silks and tassels, light choosing the lowliest, palest blue gristle for lavishing. I wait at a height and from afar, up here on the telephone wire, with what looks like a hunch-shouldered burden. Below, the red coils of spilled guts gather dust on ground. Such a red and its steam in the cold gets to be *shock*—and *riches.* Any red interruption on asphalt, on hillside, at dune's edge—*shock,* and not a strewn thing, not waste. Not a mess. Plump entrails crusting with sage and dirt tighten in sun: piercing that is an undersung moment, filled with a tender resistance, a sweetness, slick curves and tangles to dip into, tear, stretch, snap, and swallow.

The problem with coming back as a buzzard is the notion of *coming back.*

I can't believe in the coming-back.

Sure I play the dinnertime game, everyone identifying their animal-soul, the one they choose to reveal their best nature, the one, when the time comes, they hope fate will award them: strong eagle! smart dolphin! joyful golden retriever! But there's the issue of where I'd have to go first, in order to make a return. And the idea of things I did or failed to do in a lifetime fixing the terms of my return—and the keeping of records, and just who's totting it up. As soon as I imagine returning anew (brave stallion-reward, dung-beetle reproach) I lose heart. It's too easy.

Anyway, I already think like a buzzard.

The times I forget my child, most powerfully marked by the moments that follow, in which I abruptly remember him again, with sharp breath, disturbed at the oversight—those times are evidence enough of my fall into reverie, into the all that is set, unbidden, before me: inclinations gone to full folds, bone-shaded hollows, easings and slouchings, taut ridges, matched dips, cupped small of the back, back of the neck, the ever-giving body—yes, I take what's set before me. So much feels hosted—and fleet. I chew a little koan: all things go/always more where that came from.

I already know the buzzard.

That the world calls me to hissing and grunting, that I am given a nose for decay's weird sweetness, that I am arranged in a broken-winged pose to dry feathers and bake off mites in the sun, that I love the wait, that I have my turn, that no one wants my job so I go on being needed—I have my human equivalences for these.

The Lustres

I am, I admit, daunted here. Set upon by impossibility, which is both my subject and predicament. My method, then, will be the standard proceeding-in-the-face-of variety. I'll call some point "beginning" and begin. This state, right now, is coiled up like a fiddlehead fern, so bright-green, fresh, lemony, cochlear I cannot bring myself to pick/wash/steam it just yet. This moment, folded into itself, is resting so tenderly I find it hard to get going—in just the same way I cannot bring myself to make a fist with one hand while touching the yielding velvet of an earlobe with the other. Or to bite down hard on pearled barley or luminous beads of tapioca. At the farmers' market, it's the shiver of apricots, their thinnest bitter-honey skin, the speed at which the over-softness will set in, right then, right that minute if mishandled.

I have ways to manage and even enjoy the subacute rise in anxiety. The adjustments, once the words are set in motion, the circling, the backtracking, the proper dimming of lights save me.

I call upon the partial.

It is the partial I believe in, twilit and salvaged as any childhood god. Scraps and spots, moments and lustres passing and glimpsed sidelong.

I remind myself that starting anticipates a geography. A moment seeks a shape and claims *here* (bedroom window, perfume bottle) as its wobbly launch. And it is somewhere in *here* that the unsayable is lodged. How to speak of it is my problem, my subject, rolled between thumb and index finger like a bead of wool. I worry it and it soothes. Very early, I embarked on this task in its simplest form, by unspooling words: I'd hold one in my mouth and repeat it over and over, letting incantation mow down sense, so the phonemes marked a spot, trampled the ground, lit a fire and purified themselves into rote, risen things. I'd let a single, drossy word dissolve on my tongue, little plosives (*pepper*), or breathy sibilants (*citizen*) until a brief pulp of sustenance formed, a slurry juice where a word once was, and from there I could start building back meaning.

Of course I believe, still, that words harbor side streets with surprise spills of bougainvillea come upon, low stone walls and chickens whitening briefly the chinks between stones. Stiff cough and broom-swipe in a courtyard, low easy talk, internal doors slamming. I give no casual access to my city here, but think it will—I hope this impulse to speak of it will—lift lightly and settle over you, offer some sensibility, some original atmosphere. I want some lisp, a recognizable accent to surface, to catch and welcome you in.

Starting, I hoard, palm, pocket the impulse. Starting, I think: back it up, slow it down. Delay with just a little more history here.

Early on I knew. I'd suspected for some time, but then: in one of my father's art books, there was Magritte's curvy pipe, titled "Ceci n'est pas une pipe." And his four-paneled painting in which a horse is called "door," a clock is "wind," a pitcher "bird," and a valise (surprise!) is exactly—or inexactly now—"valise." So objects *were* loosely affixed to their names, and language a game we all agreed to play! This suggestion did not make me queasy; I was not chilled or moved to anger, as if toward an imposter. I was not disillusioned. I was a child. And to the word as stand-in I gave my pity and my allegiance. I extended latitude. I granted amnesty. I was *grateful* to the word for trying. So here:

I called it *Vienna,* but it wasn't Vienna-the-city. I did not know that Vienna. My grandmother and great aunt were from Augsburg, and Vienna was far, nearly 300 miles, and there were no stories about it. I could assume I'd encountered Vienna in books and arranged associations from what I'd read. But that would be inaccurate. I read no books about Vienna. (I was busy for a long time with a perfectly square little book about kids in Japan, their Children's Day, their Flower Festival. I was busy imagining my feet in wooden sandals and my waist cinched tight in a red silk kimono; the rice-powder makeup I'd be allowed to wear. The gift of an orange I'd eat with reverence. I'd have a gray kitten. And a box kite that obeyed.) I could construct other accountings of the word's first appearance, but here, now, I'm ready. It's that I took the word—*Vienna*—and matched it up with what I knew of distance and its complications. I applied *Vienna* to an elsewhere. It's one of the ways I taught myself that *elsewhere* has a shape—and that one might be, if alert, if not grabby, shone upon by its mystery.

And now that I'm in it, now that I'm committed, here are some Viennese offerings: the Long Island Railroad tracks that ran behind my Aunt Pasq's house, just beyond her small grape arbor trussed to the poles of the clothesline. Across the street, lying in bed at night in my grandmother's house, I practiced gauging the arrival of trains by the pitch of their hollow whistles sounding three towns, two towns, then one away. The profound rattle started first in the walls, rose into the vases and cups of nighttime water, and by the time it reached my chest and hummed there, it was gone.

In the moments just after the trains flew past, it was *Vienna* in the room.

And in the morning, pairing that rush with the feel of an ampoule breaking and spreading its pinks and golds, it was a new Vienna, a Vienna again and again to enter, through the door of noise (trains every ten minutes) or the buttery window, south-facing, and catching and holding the fresh-poured light.

More?

In truth, I have a ready list:

I learned the word *bower* for an intimacy I trace to a scene atop an enameled pillbox, given to me by Madame Lulu, visiting from South Africa. She ran an orphanage for Jewish refugees, and we knew the grown-up orphan who was my parents' friend, David. On the pillbox, in blue and white, a seated peasant girl and standing peasant boy inclined together in a tondo of love amid hills and a far-off, blurry castle. Their heads touched, their eyes met on the empty basket in her lap and the bouquet in his hand hung just a wisp, a breath of white away from it. Sometimes I'd take a break from the scene and flick the golden lips of the clasp apart, open the box, and touch my tongue to the fine powder left there by Madame's pills—tiny saccharine tablets for her tea—then snap the box shut and ride the wisp all the way down to the girl's lap, and fast up to the distant castle.

Contradict broke apart, escaped like a gas and entered the air precisely here: climbing a rough, wooden ladder into the loft where everyone slept. I am at my parents' friends' farm, 1970, in northern New Hampshire, a safe home for conscientious objectors en route to Canada. I'm on rung three or four following the younger brother up; he's got a hole in his thick, wool sock, the hems of his jeans are caked and dirty after milking the cow and his brother's voice is trailing us both: "Jamie, don't *contradict* me." I knew at once the word to be a borrowed one. It was clumsy, too big and too new in the older boy's mouth, but it worked, the tumblers of *contra-* and *-dict* clicking hard, the loft air both dusty and cold, the anger above and below me, pressing. I said nothing, alone with the word. Lobbed, slammed, unbuckled—*contradict*—their terrible father's word, cribbed, choric, refracted—now *theirs*. And by stealth, mine as well. Mine, since I grabbed it as soon as it fell. Over and over I turned the word, then under my breath I exploded the pieces in all directions: *contra/ry, contra/st, contra/ption, contra/ct. . . .*

And years later, traveling through France, what *are* those things, I wondered from the train—little *huts* dotting the fields?—until I could sift back for the word: *haystacks!* So changed was the word by those new forms, I almost couldn't locate it. So here were the great tufted haystacks of France! *Monet's* haystacks! And *then* I had the absolute golds and granite blues to attend the word, hold its cape, to polish and serve! Oh hidden, exoteric pinks and ochers—and *still* I could hardly get the word to fit! How these forms broke from the Midwestern haystacks I knew, baled and wired, or rolled in plastic in my little college town. How these reconstituted the tufted, leaning sheaves the poets of the Lake District so loved.

All my life, my words worked hard. Stood up to.

Withstood and understood.

I have tried to keep them safe.

By the time I'd learned *sublime*, I'd already seen its chased grays and lit hurricane greens in the Hudson River School painters' skies (*firmament*!), those parlous heights brightening to revelatory, those gorges blackly, mossily seducing. I'd already read Keats' "The Eve of St. Agnes," and to that, too—though I apologize now for the taxonomy of purples I made of it in high school English class—I retroactively applied the word.

My Aunt Pasq made us Easter bread—dense, yeasty and saltless with a hard-boiled egg, shell and all, held fast in its braided center. My grandmother grew tomatoes in her backyard. I had, for "sublime," the words *bread* and *tomato*. I had the phrase "go pick a nice tomato for dinner." And once out there, alone with my task, I had all to myself the teetering six o'clock light, the peeling and tender, pink-skinned birch shadowing the grass and me, the fence hung with flower boxes we'd watered just that morning, the fence keeping back the weedy graveyard on one side of the house and grocery's parking lot on the other. That is, *sublime* was all around—loose, though, and rampant, unaffiliated with the word for it. Even now I say it sotto voce, preferring *egg, bread, tomato, birch, wet fence.*

(I can go on, though—it's quiet enough, it's still dark out this early and I'll quickly remake that semi's air horn into a far-off train whistle. So, shhhh.) We made our beds in the morning. We did not throw clothes on the floor, nor did we put our shoes on the couches. Nothing ripped was worn in my grandmother's house. There was sherry sometimes. Some bathroom nymphs. Melon balls in translucent green bowls. A glass butter dish. We had cream in that house, whipped, with crushed pineapples and spread between layers of airy-wet cake for birthdays. On the mantle, a pair of porcelain doves. There was a vase with blue and gold trim and a Schubert-era, rosy, coiffed woman who watched you over her very bare shoulder as you took a bath. There were French perfume bottles on a mirrored tray, each with a dram of valuable scent gone brown and syrupy at the bottom. There was a vanity table, though we never said "vanity," and neither did we have the words "highboy," "chifforobe," "antimacassar." We had *shelf* and *cabinet* and *slip cover.* In jewelry boxes, a few *good things* sat apart from the more spectacular rhinestones and mod, white, baubly stuff. In the living room, me in a red velvet frame, my sister in gold. My mother as a bouncy kid, also framed on my grandmother's dresser. My mother in college, with a short cap of hair, luminous with who knows what pleasure and sadness contained beneath the decorum and perfect, sourceless light.

And always, just outside my great aunt's bedroom window (not the train-facing window, mine when I stayed there, but the south-facing one)—ah, there it was. Jump down from the high and strictly made bed, step onto the braided rug, move to the window, and there, you could touch it, open onto *Vienna*. But I never touched it. I never stepped to the window at all. I gazed unfocused, sometimes at it, sometimes through it, from the distance of my bed: *Vienna*, the way dancers locate a still point, for balance as they spin.

Vienna was far and the light superb. Viennese figures slipped room to room as my father drove over the Williamsburg Bridge at night and I strained to see into the lit apartments, holding, not speaking the word until sleep overcame. Vienna was proper, but earthy-proper. Ever-gracious. (And "whipped cream" there was "schlag," as it was in our house.) It was something *fine*. Of my son as a baby, my grandmother would say "such fine features," and *fine* meant not *delicate,* but well-done. Well-made. That care was taken, attention lavished. That he was *chosen* for lavishing. Though by whom was not the question. Certainly not by me, for I did not build his bones. By whom? There was no word for this, and if there was, and if it was *God*, it was not spoken in our house. We had, instead, beauty and graciousness. We had "Things worth doing are worth doing well." For their own sake. Which might have been pride but felt, though I wouldn't have said it this way, more *holy* than that. We had, in German, "steal with your eyes" (for my grandmother and her sister gathered their smarts and a suitcase each, and without a word of English, left Augsburg, for good, for New York in their teens.)

There was ease in Vienna and composure more than money could buy. Vienna never touched money. Not a thing there was tacky. Goodness inhered in Vienna—but it was not a hard, marbly good, in sharp focus. It was precisely out of focus. Vienna was en route, like a star, here-but-in-transit, gone even as I experienced it, even as I breathed its proto-weather and spoke the only thing I knew to call it—and only to myself. On the way to Vienna you might be attended by the smell of bacon and buttered rye toast, or by coffee cups clinking in the sink's wash water. Vienna awaited you again, afternoons—late afternoon, the light angled and private if you wandered upstairs to poke around in a jewelry box or closet or look at some postcards written in elegant Deutsche Schrift or open a drawer and sniff at the lavender sachets or splash on some Echt Kölnisch Wasser No. 4711—or there it was, just before bed, or very early morning, those points of transfer where you could, for a moment, get closer to your stop, anticipate the slowing, the force of braking pressing you harder into your seat: *Vienna.*

A word is a way to speak about something that really, in truth, no word can touch.

A word is, just for a moment, what arriving might be like—before *there* slips into *here*. And *here* goes in earnest search of another *elsewhere.*

One summer night, when I was six and put to bed while the sun still shone and the game in the street went on without me, I thought to myself, framing it up, "The world is going on without me." I refused to have the shades drawn, preferring to suffer the full extent of my exile, so the sun blared through the sheer curtains, and there was the game being played and played, way out there in the street, and soon that thought became a globe I both rode and saw myself riding; I reassembled myself upright in Japan—which just minutes ago had been under me—and, dizzier now as the whole earth kept going (how *could* it, without me!) I held on. *Champ, the neighbor's dog was barking* (I narrated), *bringing on dusk* (very proud of "dusk") and I could picture Big Jim (way younger than me, no fair!) with his folded jelly-bread sandwich, his skinny, smudgy sister, Louise, some older neighbor kids come for the game, their voices grading into the thunk of balls bouncing off garage doors as I walked in the new light of Japan in my wooden sandals and it was Children's Day or Bon Festival, and how did I come to be *here* and *me*, when I could be anyone, anywhere—oh *Japan,* my word, my key, my globe!

Vienna. Japan. My slantwise places. My bidding, my practice. I cannot, as some do to prove a point, turn a word's insufficiency into a brightening din or dangerous jumble of shards. I don't dream of collecting bits for mosaics, or cutting and pasting pretty cascades. I don't stand little ciphers on their heads, fracture and sample, or rearrange. I keep in mind a belief (how Old World this is, what a peasant I am): who knows a word is girded round with silence finds a way to realms. Behind a stone wall is a garden in brief plenitude; from the train, a blurry arbor and a woman in a housedress flash between chinks of green. From bed, another country lifts into being.

So many have come before me, come up against, come close to the task, hands in it, giving it their very best try. So many in their acute circumstances. The imperiled and delphic Ida Fink, writing after the war:

> I want to talk about a certain time not measured in months and years. For so long I have wanted to talk about this time, and not in a way I will talk about it now, not just about this one scrap of time. I wanted to, but I couldn't, I didn't know how. I was afraid, too, that this second time, which is measured in months and years, had buried the other time under a layer of years, that this second time had crushed the first and destroyed it within me. But no. Today, digging around in the ruins of memory, I found it fresh and untouched by forgetfulness. This time was measured not in months but in a word—we no longer said "in the beautiful month of May" but "after the first 'action' . . ."

And recklessly, near speechlessness, so near the snow-blind heart-of-it-all, and sidling close, Paul Celan writes, of words:

You prayer–, you blasphemy–, you
prayer—sharp knives
of my
silence . . .

and then, that last line (how does he do it, fly both earth- and skyward at once?)—

You crutch, you wing.

How can I say this—about sunlight, early morning, the path from kitchen to porch in my grandmother's house—except here, in the company of others who have acknowledged the impossibility of saying and press on. Those who have believed in the partial and particulate matter of words. *Lustres,* those prickly-bright sensations Emerson said he read for. Virginia Woolf's "moment of being," her pattern "behind the cotton wool" admitting the scaffolding that upholds, that is upholding (still, for me) that "invisible part of my life as a child." "I was conscious—if only at a distance," she writes, "that I should in time explain it . . . I was looking at the flower bed by the front door . . . 'That is the whole' I said. I was looking at a plant with a spread of leaves; and it seemed suddenly plain that the flower itself was a part of the earth; that a ring enclosed what was the flower; and that was the real flower, part earth, part flower. It was a thought I put away as being likely to be very useful to me later."

Later, when the words could *help* somehow.

Somehow they help:

When I came downstairs, summer mornings in my grandmother and great aunt's house, I'd step into the sunlight as it parqueted the floor, I was of sun as it slipped through the lace curtains, the windows were open, grass-scenting the house, and there, very heavy and bent on its stem, was an enormous yellow, or red, or sometimes, best of all, coral rose in a crystal vase, a rose my grandmother had picked earlier than morning. *Thus day by day my sympathies increased/and thus the range of visible things/grew dear to me: already I began/to love the sun* was years away, but in this way, the fitting room/dining room (my grandmother and great aunt were tailors to the wealthy) collected light, and the light spread across the wooden work-and-dining table—brown-padded to protect its diningness from ferocious pinking shears and razors and dusty marking chalk. I'd pass the (visible and dear) black-and-white patterned couch where the customers sat and waited for fittings, the always-chilly black tile of the sewing room floor (where one could easily see the dropped silver pins and magnet them up at the end of the day, if given the task), and make my way out to the porch where we ate summer breakfasts. The jalousied glass blinds were already opened, early, by my aunt, the slats turned with big, flat keys, the bamboo shades were rolled up, the table set with thin slices of crumbly yeast cake, juice in small glasses, bacon on plates, the paper read and refolded. The greeting *good morning*, not formal exactly, but a form to be followed, consistent, and yes, spent lavishly on a child. And expected to be banked and spent in return (I was not to say "hi" in the morning.) The newspaper coupons clipped, the napkins stacked in the wooden flip-top box. Hot pink packets of Sweet'N Low (garish, unlikely) loose on a tray and then, years later, in a Lucite container (also wrong, but bought for them at a school fair for Christmas). The walk through the house after coming downstairs, from kitchen, through fitting/dining room to porch, contained the barest shifts of atmosphere—the way crossing a border makes a trip into a journey. Makes it *undertaken*. Thus I learned to travel in a very small space. I learned distance contained. Walking to breakfast, I slowed like a train, maneuvering onto a new set of tracks. I navigated by way of hem-markers and mirrors. I moved between a line of black and gold Singers, each with its filigree treadle, and racks of hanging gowns to be pressed. I walked and checked the bags of scraps for anything new—a scandalous fur collar, bone buttons, "good things." We had the phrases "your good coat" and "take off your good shoes if you're playing outside." The things of a day were hued and graded, and moved from *house* dresses to *everyday* pocketbooks (in seasonal bone, black or white) and finally onto *good* coats. For kids, the progression went *play-school-dress* clothes. Later there were *evening* clothes. Much later, the customers' hand-me-down gowns, refitted for my sister and me, in case of *an affair.*

———

Here with me now are those who know that to set down words is to give to all things a partial face. Who consider the partial not merely insufficient or wanting—because to indulge such a notion would mean no work at all would get done, nothing at all would be made. For Whitman, so seemingly at ease with abundance, the anxiety of sitting down to it, the question of finding a shape for, a precision, was indeed a weight and a trial. Of a bloody night-battle, one of the war's last, at Chancellorsville, MD, in May 1863, he writes:

> Who paint the scene, the sudden partial panic of the afternoon, at dusk? Who paint the irrepressible advance of the second division of the Third corps . . . ? Who show what moves there in the shadows, fluid and firm . . . Of scenes like these, I say, who writes—whoe'er can write the story? Of many a score—aye, thousands, north and south, of unwrit heroes, unknown heroisms, incredible, impromptu, first-class desperations—who tells?

Who? Well, *he* tells—with all the terrible partiality he can countenance—what he sees, all he knows to be lost, too fast gone and unsung:

> C.H.L., 145th, Pennsylvania, lies in bed six with jaundice and erysipelas; also wounded; stomach easily nau-

> seated; bring him some oranges, also a little tart jelly. . .
> I go around from one case to another. I do not see that
> I do much good to these wounded and dying; but I cannot leave them.

And though he does record degenerate scene upon scene ("One of the officers had his feet pinn'd firmly to the ground by bayonets stuck through them . . ."), of the "real war," he says, "Its interior history will not only never be written–its practicality, minutiae of deeds and passions, will never even be suggested."

For pages and pages–and indeed the war takes up most of his unruly autobiography–Whitman offers objects touched and seen, the actual stamps, books, pens, coins, and sweets presented at the bedsides of the young soldiers he nursed and wrote letters for, and loved. He culls from his "blood-smutch'd little note-books" everything–wounds, letters, recipes, gestures, scents, finals words–to piece together "the most wayward, spontaneous, fragmentary book ever printed." He writes of the entire precise and compulsive endeavor, "I wish I could convey to the reader the associations that attach to these soil'd and creas'd livraisons, each composed of a sheet or two of paper, folded small to carry in the pocket, and fastened with a pin."

He means, I think, that he does what he can.

One night, at college (rural Ohio, attic room, wobbly desk, warm circle of light from a yellow tin lamp), Wordsworth cast forth into the unsayable in a way wholly recognizable to me. There, on a *cataract*, he might have called it, or *promontory* of recognition, I circled his definitions of those glowing lozenges of memory, those palpating areas (gone hazy at times with Romantic abstraction–virtue! imagination!) and next to his surprisingly plain phrase, "spots of time," wrote my very own simple "yes":

> There are in our existence spots of time
> Which with distinct preeminence retain
> A fructifying virtue, whence, depressed
> By trivial occupations and the round
> Of ordinary intercourse, our minds—
> Especially the imaginative power—
> Are nourished and invisibly repaired;
> Such moments chiefly seem to have their date
> In our first childhood.

Spots of time: how the words for it cleared a space for future field notes on the subject: Italy. Outskirts of Rome, en route to Minturno. I'm twenty. Friends and I are driving fast in a sports car and stop along a strip of beach to stretch, and so I can take my shoes off and touch the Mediterranean for the first time. I'd been, just moments ago, tasting a column of salty wind blowing in from the front window; opening and closing my eyes in the force of it, turning my head to catch the whistle first in one ear, then the other, pausing the sound by turning away, licking my lips to gather the salt. It is out of this private quietude that I step onto the sand and remove my shoes and walk until the big rock I see, no, the five rocks clustered, then eight come into focus, and closer, closer, they're rounding and shading until they become—is this to be expected in Italy?—drowned and bloated pigs. Bleached and swelling in the sun, stink complicating the sweet, feathery heat of October. *Pigs.* The sky is blue and cloudless and the car seats hot when we return. We're full of the scene and can't stop talking. Our pockets are lined with it, filled with sun-edged coins to keep and trade and spend for years. "Remember the pigs?" we still say when we meet, now more than two decades later.

To mark an occasion with the available props—the word *spot,* the word *scraps.* And *moments*—how meager!

But it's this, or no marking at all takes place. With no words, the occasion is gone.

Yes, words are brief, partial, unlikely, stark.

Styptic. Wanting.

Vienna. Japan. Sublime. Bower. Pigs.

Reader, forgive them all.

"Poetry Is a Satisfying of the Desire for Resemblance" (Theme & Variations)

There was the eye socket, cranium, jaw, and at the jaw's hinge, a darkened spot where muscles and tendons would gather in. There, where I stopped, were the bones of a mouth, base around which sensation assembled, arc and dip where joy would mass, interest tighten, a grin inscribe. It was a small animal's head tilted up (in sun, early fall, the leaves translucing, drying and brightening) inclined toward the three-note call of a bird. Right there, regarding the call, head back and locating the mark, not in danger or hunger, let's say *a raccoon* settled into the grass to find a sonorous point in the blue and unencumbered sky, leaves dipping into the picture and shirring, not orange/red/yellow (though they were), for the photoreceptors in a raccoon are differently keyed than ours, and *its* sky would gray out, *its* tree putty into a blunter thing from whence the call issued.

It's not that I mean to animate the world according to my whims or a lordly perspective, or that I'm bent on assigning virtues, human ones, and sowing them widely among all beings, so I might feel at home everywhere, always. I don't mean to collapse all that is between the raccoon and me, force kinship, Lia-fy any creature.

It's just that here, today, with a quantum of sadness settling in (*sadness*, not grief with its solid occasions) and a quantum of something else buoyant and lithe, I looked down (perfect skull) and then up (blue sky with birdcall) and the loop of perception closed, countervailed any singular mood, and I was less alone.

Such a feeling comes on *in waves* and one *goes under* I can say, since I grew up near the ocean knowing the excruciations of tides—not because it's *easy* to say "wave" for sadness and its workings. I wouldn't do that, not here, not now; it's more that I know very well, in a familiar way, the species of force that, without intention, draws one in, and pushes one out again, scoured and worn. Waves plunge, overpower, rash the shore, rake it. Waves sift, wrinkle and breathe. *Steed*, I learned later, for intense, white-foamed things (*The trampling steed, with gold and purple trapt/ Chawing the foamy bit, there fiercely stood*) so yes, there are many angles to consider—sound, for instance, the tight squeeze of those *e*'s and the *o*'s invitation—when noting that waves work well on behalf of layered-up moments.

(And, I should add, when I first saw Rembrandt's waves—it was *The Abduction of Europa*, in a book at home—new shades of sensation were affirmed; I could find, after that, in puffs of real sea foam at any local beach on Long Island, the bull who was Zeus, bearing Europa fast away, Europa seeming up for the ride, muscled and ready, balanced and whole, borne over waves on another's will, the roiling, darkening sea inviting, the Europa in me RSVP-ing, all I was leaving, and all I'd be finding churning together, suggesting. . . .)

I so loved the ocean as a child that I had to be dragged out when it was time to go home. If you've grown up with waves, you come to learn that they don't knock you down as much as allow you various decisions about staying upright, show you've chosen to stay in their path, try your luck, pit your strength. And though we say "a wave knocked me down," it's not that waves care. They're as rote as heartbeats. *Down*, though, draws the eye—

because Dante's geography promises you'll find your very own species among the fallen. *Down* because Lucifer, who once tended light, fell away from the light, and now lives below us. Because down is where we go for essentials, where we seek the authentic by way of the thoroughgoing need to come clean: *Pull down thy vanity, I say pull down/Learn of the green world what can be thy place. . . .*

So let me confirm: when the bird call came, I was looking down. And there was the skull. Surrounding it was a sensation, and above, a sky very deeply blue. Then it happened: the picture got bigger: the skull was, I saw, not a skull at all, but a weathered mushroom, eaten back, or worn away. The whites and creams, the holes for cords, the holes like sockets and the slendering snout—all turned back to gills/stem/cap; there was the shift from bone to mushroom, a rising from solid and going to pith, rigidity softening into flesh.

In the space a mushroom now held, for full, long seconds, a skull had been.

That pinned me to the afternoon.

To concentrate a skull up from a mushroom . . . but no, that's not it. It went very fast. It was vaster than any conscious thought. To be of a moment that folds up distance, *finds* no distance between mushroom and skull, allows *skull* from the first—though there was a patch of new mushrooms right there, shining, fat, rampant, creamy, just-sprung. To be part of a mind that flies past the known (until finally, the cues come on hard: all those days of good, soaking rain, the fast greening of lawns, everything sprouting and shooting like crazy), to be part of *an order, a whole, a knowledge, that which arranged the rendezvous:* at that tufty spot on my neighbor's grass, with an airy/oceanic blue sky above, mushroom met skull, the resemblance bloomed and extended me. Right into the heart of the afternoon.

Such resemblances get made in other ways, too:

Once I spread my fingers and looked at the spot where thumb and wrist meet, and in that depression saw *soup plate* (what my grandmother called any shallow bowl, and hers were cream-colored, low-fire clay ones, with flat rims of green—how suddenly that comes back to me!) then *crux of a tree for holding rainwater; a hammock; a nest.* I saw the imprint of another thumb's work—I'm not saying God's (that's nothing I'd say) just where an actual thumb would have worked, should I have been clay. I considered, too, how other thumbs *have* worked, right there in that spot, but for pleasure, roiling oceans, vastly, in me—

Once there was a wound I was tending. *How high that highest candle lights the dark* I spoke in my head, to steady it all, because the tending made me woozy. The wound was a taper that went far in and down. It involved the colors of a candle flame—what the body chooses for regeneration, chooses to light its dark passages with!—and this was a perilous passage. For a while the light moved like a tide, receding then overtaking the shore, the sweet, cool sand that was the good skin. The known world was there, beside the reds, fatty yellows, off-whites—colors by themselves not at all unpleasant, but on the small island that was the wound, threatening. A wound grows together from underneath first, the inner muscles knitting up, and the surface is the last to close. It all cinched slowly back together (with oxygen treatments, medicine, rest) regained the right pinkness, as the whole body did, regained, as we'd say, the *rosy blush of health.* And indeed, when it healed, it *looked like* a rose, was *roseate,* a furled, tight bud of a scar, and one day, exactly that—*rose*—was my first thought and not "wound."

And once, very suddenly one afternoon last spring, I saw that the apple tree outside my window had grown into the only spot of sun available to it. And so, because there are pines around it, thick, tall ones, and the sunlight is meager and hard to come by, the apple tree is terribly bent, sway-backed and leaning.

A thing grows into the light available to it.

This is not just a metaphor.

And that a mushroom is also a skull, is not a trick of sight alone.

Against "Gunmetal"

June. Cape May, NJ. Boardwalk.
Rain coming harder. People hurrying. People jumping boardwalk puddles with bright sand-centers. Avoiding the spume of passing cars. Ingraytensifying the soft dunes with neon rain gear, all the ponchos calm and isoscelate, then blown scalene in wind. Now it's more to watch, the dodging and pitching. More, maybe, "fun." Of interest. "Human interest," because rain alters people in unexpected ways. And the unexpected makes people so human.

Remember that.

Out there on the boardwalk, they're absolutely dedicated to being human, and though not one of them has a choice, many variations come forth. All the ways are recognizable, but some are more precise in cast and tensity, saturation and value, and take patience to see and to name.

Outside's thunderclap, its tonnage and stipple. The toilet in the room above's flush. Extended, deepening thunder sounds. The picture window's darkening glaze. Except for the mother with her hood pulled tight, a sporty family neverminding the rain, laughing, carrying big, wet cups of coffee. A runner tendon-stretching, braced against a stop sign. An old-salt type in a long, yellow slicker waving to someone, or directing the deluge. More cars than usual heading north to the parkway, as goes the decision through many heads at once to leave the shore earlier than planned. Methods of resignation abound: one on a gearless soft-seater pitches into the weather headfirst, a sack of oranges hooked on his arm. Four pedaling a surrey remain committed to their rented hour. The sky brightens. The clouds shift. The cars slow and their numbers decrease. Runners come out, had they ducked under awnings. Outside our window, a gaslamp-style streetlight's on; it must have self-lit at the first hint of dark. Walkers wearing long sleeves and sweatshirts, who must have tested (head out a window, arm out a door) the temperature before emerging.

Various pitched rumbles, filling, ablating. A rough sound, that otherwise might be silk tearing, but for now is tires parting puddles. All headlights on. Sky darkening again. Those choosing to be out or having been caught, somewhere on those bodies in the noisy rain: shiny, cicatricial spots of damp. Wet shoulders where clothing is sticking. Abrasions on ankles where sockless shoes rub. Itchy tags. Rings of sweat. Objectwise, sunglasses in bags or hooked at collars. Loose, jangling change. Newspapers rolled and stuffed in back pockets. Some lightning now, but candescent, not the sky-ripping variety. Some darkness lifting at the horizon, baring a strip between sea and sky, like a hem rising over a sock.

Now the umbrellas, now that the walkers have figured it out: *rainy* not rain. Dark as any November day, late in the afternoon. Blue turned to its compounds and alloys, its milkier elements, whitened and hardened. On Beach Drive, the activity increases: doppler riffs. Gutters surging. Thunder yanked like special-effect sheets of aluminum, behind the scenes. A jogger who can't economize movements, whose legs seem strapped on and lack propulsion, whose elbows angle too far from his body, seems wetter than others. Bending in wind, heavy with rain, some hardy beach roses suggest a boat tethered and scuffed against unseen pilings.

One species of sleeping person can sense rain and somehow knows to stay abed, undisturbed in their summer rental, up and down the beach. An announcement such as this won't jar them: *May I have your attention, please. Lightning is on the beachfront. Lightning is* on *the beachfront. Clear the area for your safety*. It sounds not at all canned: the voice of a real and excited someone, red-faced, soaked and bringing the news. At the horizon where ocean meets sky, a mist congests and erases perspective. Rain threshes the sand. The sky darkens further. The sky turns, *toward* or *into*. The sky now. The sky is—what *is* the shade, gradient, hue, tint I'm seeing? The ________ sky. That sense of searching, fingertips tapping, calling forth terms. Sifting, anticipating: the *something* sky. Something. Something pushes in. It draws up to full height.

It blots out any other sky, *gunmetal,* does.

How irksome. *Gunmetal.* What a cliché.

Strike me down if I use it again. If I don't, right now, erase this method by which we impart, those of us who know nothing about guns, drama to a sky, pressure to a scene, hardness, knowhow, coldness to a description, glad for its hint of treachery, its sidelong, thanatotic meanness.

Why erase, though? Why deny the relief of a shared, common phrase—novelistically charged, not the worst imaginable? *You* know gunmetal and *I* know gunmetal: why not meet there? Pretend it's a bar of the same cool name, "Gunmetal's" (brushed steel, understated track lighting) and relax, converse, affirm each other's positions on many Big (or breezy and minor) Life Issues. Since I had nowhere to go this evening and you were free, and isn't that better than staying home? Even if I know where the conversation's headed? And really, you're perfectly decent company, *you* aren't at fault. But after an evening like this, I'm way more antsy and hardly refreshed, since I'm not at all changed or challenged or stretched. And neither are you.

And yes, the *coldness* of a gun pertains. A gun is, when you first hold it, very cold, and way heavier than you'd think—say a .22, hitched right up against the shoulder. At least the one I shot weighed more than I expected, made as it was, of . . . I don't know what. Gunmetal, I guess. I hardly have anyone to ask about this. One strictly seasonal pheasant-hunting friend, who will answer modestly and not say one thing beyond what he knows. Another who fought in the Iran-Iraq war, and though that's long ago now for him, I hesitate. Because maybe it's not so long ago, the way rogue scenes slide in when you're making a sandwich, washing your hair, touching your sleeping child's face. . . . Also, I've seen that tree, in the photo in his living room, the tree he's standing so uprightly next to (he in his uniform, and both so thin they look related) and *something* came just before the photo and *something* happened just after it, to the side of the tree, or behind it—it's that the tree's starkness is a point of reference. There is, I think, a lot more he knows, for example, on the subject of grenades, that I don't want to ask about either, there being no "grenade blue" I'm harrying here. Though there's a sky for that, too. A misty tint, a haze indicating surprise detonation, rain turned to hail, very suddenly.

But I want to know what "gunmetal" means, and found the perfect guy to ask, a friend of a friend, a gunmaker out west, who's currently working on a matchlock from 1510 ("older than all my friends combined" he says.)

My questions, of course, are embarrassingly basic.

And yes, I do need to start at the beginning.

Jim writes: *Glocks are made of plastic with metal inserts in the receiver or frame (the part you hang on to), the slide and barrel are metal and the color is determined by the options you choose*. (He's seen pink as well as sky-blue ones). *The basic metal a .22 is made of varies but it is always shiny silver, what we in the trade call "in the white." This reflects that it has not been colored or coated yet. The coloring (whether it be bluing, Frenching, coating or browning) is put there to keep the metal from corroding or oxidizing in an undesirable manner. "Gunmetal" as a color is usually a gray, more technically called "French gray." Think of the dark ash on charcoal, only shiny.*

The shiniest guns would be chrome or nickel-plated, the blackest ones would be the black epoxy-coated; black chrome is black beyond belief, but is shiny like a mirror. These coatings can be applied to any firearm. I have examples of almost anything you would like . . .

Almost anything I would like . . . as, too, this sky is variously compounded, concussive, concupiscent, and oh, could be layered with names transfinitely: it's the rivery color a silver spoon turns when held in a flame. It's the color of a well-used plumber's wrench. A perfectly battered railroad tie. I try on: *A burnt-spoon sky. Below a sky where we sat down, under wrench-colored clouds. Before the sky opened and a rain as hard as railroad ties fell.* . . . It's the color of a cataract (which, very like "promontory," is not much in use, ever-nailed as they are to the nineteenth century, provenance of the Lake District poets). It's a kinked intestine-gone-bloodless-pale sky. Translucent, unfeathered, fallen-chick silver. Powdered zinc. Stripped olive pit. Dirty-kid water in a porcelain tub. Farinaceous. Clayey. Grime in pressed tin. So why *gunmetal*? If it's something about the act of smithing, why not things from the worlds of cooper, tinker, wainwright, glazier? I suppose the throwback quality's engaging—the forging, the shine, the bluing, blacking and browning—but mostly, I think, it's rugged and hip to suggest with this phrase you know something about guns; enough at least to toss likeness around. You have to like a likeness to toss it (note: kids running, jostling, outshouting each other as they race to a car will call *"shotgun!"* not *"side saddle*!" not *"the seat next to my mom"*).

If you're really set on naming a sky by way of armaments, try a breech-loading carbine's pencilly softness, or another from the Civil War (see the excellent display at the Gettysburg Visitor Center), a Harper's Ferry musket whose mottling looks like winter rain. Try a cannon's smoothbore, or case shot, the spherical or precisely penile munitions, pocked, blackened and smutted by all the ways they ruined a body, rolled, muddied and were gathered up again for duty. Try the brass coat buttons, buckles, and plates identifying cavalry, riflemen, musicians, artillery, infantry, engineers, and the tarnish spots there, *that* color, where the salts in blood wore away the finely wrought eagles, lyres and flags. A mess cup's the color of the Potomac in winter. A bayonet's black as a rasping crow. And "rust," it turns out, is a complicated blood-dew-gunsmoke amalgam.

"Battleship gray" is also a problem; consider the monstrous snout of a ship, fastened with rivets the size of plates, unyielding and lithic–does a *sky* intend to communicate this? To bear down, to invade? Can't we come up with something other than a destroyer's brutal, flat gray to signal a presence that hovers over with steady nerves and conviction? In Farsi, my friend offered *Ghamangeez*, "a saddening sky," and his wife refined it: "a sky that brings on sadness."

Okay. Now we're getting somewhere.

It's quick, gunmetal is, and efficient. I'll give it that. It speeds the scene. So you can get on to something else. It's a term that makes you feel part of a team. A baton you hold firmly and pass down the line. The way a party icebreaker works: let me introduce you to X. Now you're friends. Now the two of you can have coffee together. Then *you* introduce. To one of *your* friends. They go for a drink (you know where). Now so many of us have something in common. We're cozy. We know what the other means when we say. . . .

Skies change, thankfully, and grays complicate–unfurl, turn smoky, egressive, specular. A few hours later, the sky in Cape May has taken a turn that stymies. It stumps me. Car base coats, the flat ones, rally to help. Giorgio Morandi knew, and applied to the bottles and humble plates in his paintings a range of opacities, the soft, cool creams of unspeckled eggs, of froths and dunes. Of dusty, white Neccos (whose flavor is cinnamon, and surprisingly spicy, almost fireball hot, but muted and sweeter, so the shock spreads more evenly over the tongue, with no ping, no ache, nothing tornadic.) *This* sky is more oatmeal, ashed incense, clamshell. It's the color of shit in its calcified state, though this likeness is not much in use, alas, our palette's not very broadly accepting, and shit is not aesthetically easy; it won't stay domed. Won't stay chapeled, as it is when left alone to dry into earthy, roadside temples. Fat white gulls and snowy egrets disappear against this sky, which makes its color more erasure than presence. Ghosted. Palimpsistic.

Birds can't sink into gunmetal skies.

"Gunmetal," on the comportment family tree, is close to "steely." Steely eyes. Steely wills. Ramrod posture. (And ramrods, of course, pack down charge in muzzle loading guns; thus a body fit to load munitions, push explosives, shoulder them in, so straight and stiff, it must have been trained. *To fight, to serve and never to yield,* its motto might be. A body like that. A sky like that. Mission-bound. Singleminded.)

"Gunmetal," deployed, delivers a payload of routine. And routine is a much sought-after commodity. I get that. The best of us succumb at times. About McDonald's, for instance, the Cape May guidebook confirms, "You can't live on gourmet food alone. So it's comforting to have Mickey D's right here! There are few things in life more reliable or comforting than a Happy Meal. There is something to be said about knowing EXACTLY what you're getting EVERY TIME. No worrying if your steak is going to be cooked enough, or if the clams are bad. The only thing to worry about at McDonald's is whether to get your meal small, medium or large." "Gunmetal" as Happy Meal. It's compact, the phrase "gunmetal sky," as reliable a delivery system as any Big Mac, withtwoallbeefpattiesspecialsaucelettucecheesepicklesonionsonasesameseedbun. (How cleverly that little jingle—I can still hear the tune—indicates both precision and overabundance.) And though I won't go on with this point, research shows there are 380 seeds on each sesame-seed bun, "give or take a few."

So when I say the word to myself, for a sky's particular depth and hue, "gunmetal," which precisely means "dark gray with blue or purple tinge" (but you knew that, didn't you), a third, nictitating lid comes down and though I *see* the sky—more accurately, the real seeing stops. The little path meandering out, where I went hunting all this time for other colors the sky might be, fuzzes up. It bombs the path, "gunmetal" does. I'm trying to locate it in my body (say at the spot where clavicle and shoulder meet, where the rifle kicked hard and knocked a week-long bruise into place) so I can say the word "gunmetal" and mean it. But I don't feel it. I just join with. I fall in. I get phalanxed with the staters. Heads of Statement all start talking. All agreeing, nodding, yessing. Settling. I feel I've been given one of those ovoid bumper stickers, alerting all to my vacation spot—that mysterious "OBX" (Outer Banks Crossing, I learned at a stoplight, eye level with an SUV's bumper). Or in Cape May it's "Exit Zero." Very in-clubbish. The longer I stay in a place, the more okay the decals seem. I'm hustled in with the locals and after a while—we've been coming here for years—I begin to feel pretty local myself. Happy to be readable. Glad to be part of.

At luminous moments I have wanted to say, "How blessed I have been"—but can't. My problem is accepting a gift so weirdly, singularly bestowed: why me? why not them? I'm better with gratitude that's more diffuse: late afternoon in the middle of my life, cooking dinner, the window open, sun releasing the scent of pine floors into a solitude still and light-scoured. I'm more at home with moments beflecked with goodness, than I am with the handed-down-from-on-high kind. Things like plumbing and clean hot water, hard, tart apples and well-sharpened knives best set my gratitude in motion. "Gunmetal" would make a follower of me; using it, I'd have to say a thing I've been taught to say. Believe a thing about the sky I've been given to believe. As I'd have to take "blessed" to mean: I have been chosen, marked, held right in the center of some kind, crosshaired sight. Which is nice. But doesn't the universe also fix on falling sparrows, lend its attention to spectacular disaster, train its very steady eye on accidents, suffering, diminishment—and not intercept, help out, bless them?

I want such a sky to quiet me (not "strike me dumb"—that's a rod drawn up, enforcing awe, and one is "smote"). And I want, in that quiet, to search out my terms. And what I decide on, I want to be more than a firearm's alloy. Harder to come by. Stronger. Chromatic. I want to turn to oyster and mouse, tidepool and tin, and then tank those and reconfigure if the gray they offer is not worthy, if associations gained are not surprising, of a distance previously unreachable, and intimately roomy. Freshening and new.

Or let's not play Name That Color at all: goodbye to Keystone, Gauntlet, Cloak, Summit, Uncertain, Vast, and Repose (from the neighborhood paint store's line of gray offerings) and take up geography and spatial relations—how far, in what way, for how long did the sky lift away from sea, hunch in close, or variegate.

Or activate good old "gray" as a suffix, but hitch it to actions like *torque-*, *welter-*, and *brim-* . *Coruscate-*, *grizzle-*, *rave-*, *solder-*, *convulse-*.

Or consider that which disappears into the sky—bottlenosed dolphins that leap-because-they-can, their play, research shows, both useless and necessary—in other words, restorative. Dolphins leap because muscles want flexing, because the air at Cape May in June, is warmer than water and the change is pleasing, the shift between elements tickles them. In fact, I just learned, dolphins mate up to eight times a day—even when not in heat.

To disappear into an endless, dolphin sky.

To sift and sift and sift words–and not find. And in the face of not-finding, to not-rely-on. To turn away usual corollaries. To maybe just sit before such a gray sky and give up, until strength returns and possibilities rise. Or maybe just watching is enough. To unburden in that way. To unwind. Take it out of your pocket, your holster, that sky. Lay down your gunmetal. It's the sky buy-back program. The sky amnesty plan. Turn it in, buddy. Hand it over, right now, while you can, and you won't be charged with theft.

And now you're free to find your own term.

Street Scene

At least I can't *identify* a particular state of mind—nostalgia, say, thieving from elsewhere, or a stricter, plain yearning at work on the scene. Rather, just sort of blankly did I enter the car, start the engine (those three Hondaic chuffs before catching), and drive it into the sky.

Or (and this direction is also possible) I submerged it. Or hit a misty wall of rain from a fugitive jungle. Steered into a cloud on the lam.

I was on a familiar street, but I had to *assume* this—an epistemologically unsound move. My watch showed I'd been on the road for just minutes—a fact that helped not at all to tack down the street which had blown, which was currently falling and shushing, like a sheet of paper, unreachably far under vast, inherited furniture.

The street slipped a groove like a kid's jostled train set; the street behaved like a charcoal-sketch track laid lightly down and easily smudged. Puddles settled in dips and depressions. From a distance they were blackened blots, and then, as I neared, they silvered over. They behaved as expected, dependably tilting the sky one way, then another, rippling, wavering it. This puddle series was arranged with some logic: if splashed by tires (or the foot of a distracted college student), the puddles recommitted to new, nearby watery communities, mercury-style. The solid, central yellow line divided the street neatly; it was a nice painted line, not one of those newfangled, rubbery strips set down with a waggle where the road crew faltered, got ahead of themselves, or behind in their gluing. (How I've wanted to pry and reroll those strips, stash them in my sewing box along with coils of binding and red measuring tape! And here I comforted myself: at least I recognized the yellow thought, and pictured my blue sewing box, and home.)

But the street itself slipped free. There was no alembic click of light and shadow. A lyrical moment, highly quotable and good for rainy occasions like this—*petals on a wet black bough*—did not appear as an apparition, to affix the scene to a recognizable mood. No ensemble of clues plotted the meeting of a St. & a Dr., a Rd. & an Ave. to orient by. The street just would not, would not mean *route-to-store*, or *close-to-home*. It offered no eau-d'library-nearing (white roof the top note, crowns of maple the finish), no whiff of farmers'-market-upcoming, no low-grade-parking anxiety-flicker. No overture, prelude, or preface rounding toward anywhere stepped forth. I was driving—first principle, sure—but it could have been anywhere: Baltimore, Barcelona. No last-year-at-this-time specimen (that yellow moon slung low over IHOP) (now Enterprise Rent-a-Car) turned into a wist- or a joy- or a hurtful past moment.

As I said, I might as easily have been flying, all movement unfelt, the speed of the moment so wholly contained, the distance covered, unrecognizable. The street's singular elements were perfectly nameable—that echt yellow stripe, those newly-dribbled tar snakes filling cracks, curbs darkened with rain, fickle puddles, passing cars launching watery stars out of low spots to firmaments elsewhere—yes, the *things* of the street were nameable, but helped not at all to locate me, as when, from a plane, looking out, looking down, certain of an actual neighborhood below, the internal eye conjures up joggers pushing triangle prams, bike bells aflame in low sun—though the whole of the landscape remains a big chunky patchwork and nothing on a human scale asserts.

I knew the street to be "residential." It leafed over with well-tended trees, curbs dipped politely at corners, I could read all that, yes. Those clues registered. And so did the need to go slowly—but only as reflex, a synaptic response. I had not the sense of a specific school zone directing the downshift, or that, say, a tumbly, yellow-haired kid in favor of darting lived near.

So where was I now?

And also, *who*, is the question.

Here, into the picture (I'm slowing this down, considerably), came an old woman shuffling, assiduously not looking both ways as she crossed the street. The crossing was a big, concentrated-upon project—an endeavor which must have, earlier, as she dressed for the day, required planning and determination, the gathering of *moxie*, as someone's grandmother would've said. Or she herself would've said. And at this she'd laugh quietly: "moxie" applied to crossing a street! How silly the way age reduces us—a trip to the drugstore, across the street, *planned!* My question would be hers, too: *Who am I now? To those college kids in the new apartments—part of a tribe*? (she tries out "The Olds.") *Daft?—yes, a little-seeming, I'm sure. Harmless?—oh, all the harm done in a lifetime, now done with, and time-softened, sort of. Now (*she thinks), *I'm the person who cannot believe she was once one of them, that eye-of-the-storm, centrally pumping heart one makes of oneself when young, all confusion and terror and beauty at that age.*

This congeries of moments wasn't long lasting. Was startling, though. Microdramatic. I was trying to find my way back, or dig into the moment, there on the street. It was the sensation of trying to raise a stuck window, knowing the stubbornness to be weather, the resistance to be moisture, and that, with the right blow applied, it *would* move. I felt, too, the thought's construction shift, to a new shade of doubt—an insistent, keen, little stab: it would *have* to move, right? The moment's flat, solid resistance *would* give, would *not* behave for much longer like a colony of coral, with endless, internal, spawning lives?

For the duration of the moment, trying to locate myself on that street, trying to tack along it, I was as a foreigner. I was en route, and in a strange station, bars of the ticket window striping the clerk's lips (straining to read them), the quick clerk reciting time/track/tariff, the echoey loudspeaker announcing my train (its delay? its departure? the dining car's closed?) I felt, amid all the commotion . . . what? In such a chaos, listening for that sound-combination meaning *my train* and *my destination* (dear friends, in a foreign land *always* memorize the "from-to" construction and numbers at least to one hundred) suddenly, a fingernail shone promisingly out; it steadied me, there in my car, as it did one real and wintery afternoon years ago in Warzsawa Centralna. I was booking a couchette to Prague (do *what* with my visa—stamp? save? submit?) when that very same nail, mine own, seemed to offer answers about my journey—if only I could read its whorls and shy ridges, its dents, and the crescent of Polish dirt there collected!

So what happened?

I rolled down the window for a breath of air. At least I still knew to regard with pleasure the way the stiff, hand crank activated muscles in my shoulders and back, and sharpened and fixed my attention. The exertion felt good. The wind lifted my hair and found a way to my neck. Shifting into first gear and lurching forth caused a line of cold rain to slip from the roof through the window and onto my thigh. It soaked in and darkened to a tiny Brazil. I eased into second. Things cleared.

What *happened*?

But I've already told what happened.

All along, this has been the story of a moment.

The cross-sectioning of a moment *is* the news. That a moment anywhere—here, on a street—*does* this, is news.

Later that week, to keep the feel of that moment alive, I studied up on the construction of streets. I liked one particularly precise diagram I found, showing how different materials are layered to provide flexibility and skid resistance. Internal steel beams or meshing help a street withstand cycles of expansion and contraction. All kinds of seasonal flux is planned for. Subterranean drainage systems with rocks and sand control saturation. A formula called the California Bearing Ratio is used to calculate for appropriate loads. One cut-away showed the world of buried telephone cables, gas mains, sewer pipes and other bundled electrical stuff—all the systems collecting-from or delivering-to each nearby house its heat and waste, its light and voices. On top of the street was sketched the outline of a house. Then behind the flat house, a blue wash was meant to stand in for the sky.

When I looked up, beyond the diagram and expanse of my desk, there in the frame of the window was grass-sky-trees, in no order at all. In no order at all, it went phone line, kicked silver trash can, far steeple. It moved from green coil of hose up to far-off pink cloud.

I found, given a cross section to study, the eye hovers and slides, lingers on the most satisfying shapes, won't follow a plan but pulls in, zooms out, sharpens some things, dissolves other things. The eye disarrays the neatest sequence.

Thus in the frame of my window now: a white truck. And just to the right of it and up the steps, a porch swing with two people and a baby held close. A mailbox, a white post, a set of gray shutters. The baby in a bright red hat with a tassel (last week she was born) (the red's easy to see). A mother and father laughing and singing. And a child, already moving in and out of the scene.

Being of Two Minds

Our playing field is completely overgrown. I'm calling it a "playing field" though it was just a bare hillside with rocks we plucked and threw into a sewer grate for a game. But it was not "just," as in "inconsequential." I only mean the field was in no way official. And I mean to be neither sentimental nor nostalgic—though to say *our* field does mark it with an intimacy, I realize. To present a little history here, even if remote and sketchy, to let you know this site is charged and layered up is important, so that I might best grade into the state I am bent on exploring: being of two minds.

Passing our field, some milkweed fluff blew onto my black T-shirt and I let it stay, thinking *fuzzy-edged cloud, spun sugar halo. . . .* The day was so beautiful that I laughed, the sky so absurdly blue, June first, it seemed apologetic, a making-up-for. I laughed, and the laughter was not tinged with sadness of any kind, for the game we played was of a certain time and place. It was meant to be contained, I know this now, and looking back, the game itself was absurdly blue and lit, a respite even, like this day, in which nothing, for once, came up about *this all* going, *me* going, *everything* too soon gone. I crossed the street and saw a parked truck covered in AstroTurf with hundreds of little plastic animals hot-glued on at all angles. As I passed and looked back, I saw, hand-painted in white on the bumper: "Laughter drives the winter from human faces ha ha ha. . . ."

I was not of two minds at that moment. Instead, I laughed easily, without thought or effort. Whereas two minds *come in*. They find you. They wrestle and present cases, part waters and curtains. There can be legalese with two minds, and wranglings, and shadows vying with rays. But this was one mind—the freedom from sadness, from missing the game; the bright weather; the truck with its tailgate afterthought; and the day, or moment at least, unbeseiged.

Then, closer to home, came a yellow rose in the yard of the hands-down best gardener in the neighborhood, wet at the top of the climbing bush, bent far from the lattice, heavy and shirred on its stalk, but upright.

There is a way a flower can be frightening, and this rose was emphatically so. It was doing exactly what it was called to do at the moment, in that instant, the only moment there to receive it. Wholly in time, it was fixed to its task, with all consequence still ahead. It did not refer to Shakespeare's Sonnet 50, which earlier I had been reading: "For that same groan doth put this in his mind: / My grief lies onward and my joy behind." No. Centrally commanding as the rose was, as a heart is, it was not a scooped center posed between griefs. It was the yellowest buttercream custard *and* bowl at once. Unto itself, unhinged from time, I saw it. Not "timeless" in its beauty, but loud. It was, I think, laughing. That yellow might have been a "peal." There might have been "mirth" or "glee" in its face. The rose might have grown "on a lark" (then flown!). But not then. Not just then. It was fat and its wings were folded. Nimble and fearsome in its flight contained, its one aureate face/body/mind bent on neither staying nor going.

Really, I think there are more than two minds.

But a third, bent on settling up: that's not the state I'm after here. Not a perfectly pleasing, measured harmonic, a synthed and kindled happy medium. A balance, a stasis; form on its way toward resolve, that cant.

I think we are up to—out there—eleven dimensions.

I do not believe the earth is flat.

But I still believe in the humors. I subscribe to all that good theory, from Hippocrates on down, about the origins and travel patterns of feeling and disease, trade routes of blood and phlegm, the yellow and black biles coursing or slogging, the charts measuring consequences of overflow and congestion. I believe in the humors with their assigned temperaments, dispersed and roaring throughout the body, each with its province bounded and hued, its climate matched to the elements residing in spleen, heart, brain, and liver.

Of course when I'm driving long stretches, I still pull toward the *horizon*, numinous line that exists and doesn't exist.

And when I saw the autopsies performed, the blood therein was poppy red, red unceasingly, and no misty or frothy, clotted or blackened bilious poisons rushed forth from even the most ruined bodies.

Two minds certainly complicate one's mythopoetics.

When I read Dickinson and Whitman back to back, I am reading for the precipitous rise and fall between them. If styles are territories, I want to tack along those open ranges and consider the America that holds them both. I read as if trekking, for the recovery period in which a musculature repairs between one exertion and another. I read for the crevice that opens between the vastness on either side. And I read to fall into the gap there, to be the place where the two shadows go syncretic. Sometimes I get confused: whose shadow, whose shoulder *was* that rubbing mine?

I have *two* shoulders, I know, I know. There's a voice stationed at each—her strain, his force; the oracular and choric; whisper and yawp. And there between them, I tense and hollow out pockets in my collarbone. I make myself a harrowed place into which each abundance, each with its differing cargo falls, for one is not more dense than the other, or more weighty, ecstatic, agonal, dire.

The scission that has been made between them, I am not upholding.

———

On my way back from Poland where I lived for a year—almost two decades ago now—I sat next to an old woman on the plane. She wore a long skirt with a long-sleeved blouse, and a heavy wig with a scarf, though it was June. She was the wife of an important rabbi in New York. They had both survived the Holocaust and when she reached for her bag overhead, there were the numbers on her arm. We were talking about famous gardens we both knew in Russia, England, Poland, and France. I told her I'd never had much luck with my own, how they were always a mess, everything straying and overrunning the beds, getting out of hand and defiant. She described the gardens she'd always kept. She spoke of her roses, zinnias, dahlias, the tangle of vines netting over the fence, everything crammed in a too-small space. "You know," she said in her thick accent, "I love them all. All the weeds and flowers. I keep even the dandelions in." I remember thinking *I recognize that.* And I remember feeling shaken by the recognition, the neatness and the wildness unresolved. That she was not, could not be, discerning. I remember staring into the dirty gray weave of the seat in front of me and thinking, uneasily, *This is the only way anything will ever make sense to me.*

———

I love the friend who is slow to talk, whose composure is a hard-won grace, who works to find a rent in the persistent heavy folds and drapes that weeks–indeed whole seasons at a stretch–can be, and laughs despite.

And I love the one who stands easily close, goes rib to rib, leans in, eyes closed and sniffs and says, "We are all such animals aren't we?"

———

Two minds must state their position, as in any good debate, and fight it out:

I believe in progress and that we get better.

And I believe we inhabit a form/countenance/aspect so essential it cannot be altered. (I like the old-fashioned term "bearing," the idea of a temperament directing our actions.)

"He had two selves within him apparently, and they must learn to accommodate each other and bear reciprocal impediments. Strange, that some of us, with quick alternate vision, see beyond our infatuations, and even while we rave on the heights, behold the wide plain where our persistent self pauses and awaits us"–wrote George Eliot/Mary Ann Evans.

———

The woman with the long gray braid, who walks her grandchild to school, ought to cut her hair, I was thinking this morning. The braid is too long and limp and wispy and looks more like a baby squirrel's tail than a gathered plait, waiting to be undone and let cascade. I thought "terrible to be old"–the sinking, the shuffle, the loss of balance and ease of leaping over curbs . . . but what is it, really, I'm turning away from? Evidence of time as it rests and elaborates in the body? Time commandeering. Offshore detonations rocking the waters. That it's shameful somehow to be made helpless by time, its scouring away of the individual form one worked so hard, over a lifetime, to constitute. It is the custom of some old women to wear beige—"bone" my grandmother called it. Not "tan" or "ecru." Not "eggshell." Not "khaki"—right gear for stalking the land of bones—but "bone." As if that were a color along the spectrum. Or a charm—so that, disguised as the thing you'll surely become, the angel of death might pass safely over your house.

I hated the meekness of that color. *Let me never wear "bone"* I'd think.

I, who sit daily in front of a collection of real bones, three animals' jaw bones with rough, flat planes and holes for cords, and sockets for eyes, all flesh picked, washed, burned, eaten away.

———

I just read: "What others might have called the futility of his passion made an additional delight for his imagination. . . ." (George Eliot)

Well, no.

And yes.

Two of my oldest friends just visited, each briefly, and returned home, one to England and one to Italy. I miss them now, and in their absence, know that I will never see them enough in this lifetime.

And I also feel held by the atmosphere each so recently scented. Right there in my kitchen was the gesture of hers I'd forgotten, long elegant hand at her flushed neck, a moment of restraint before launching her point. And there, still, the sharp tooth that shows when he laughs, and the quick eye that follows the curve of a pear, reddened in one spot low on its rump. So I go back and forth. Bereft/held, bereft/held: my heavy, iambic, two-chambered work.

And yes, the eidetic moments help.

Here is a favorite sentence from *Ethan Frome*, a marvel of lightness and economy: "Once or twice in the past he had been faintly disquieted by Zenobia's way of letting things happen without seeming to remark them, and then, weeks afterward, in a casual phrase, revealing that she had all along taken her notes and drawn her inferences." By the time that first comma arrives, and then the next two which so quickly follow, the route of the whole sentence is cast. I remember my brief anxiety there—feeling the shape of the sentence forming, hoping the second part was up to the task. Of course, that part *is* up to the task—the whole sentence is clean, spare, beautifully paced. It's a hinge in the story, too; events *turn* because of this sentence, loitering intentions ripen, recrudesce at just this syntactical moment. I love this sentence because it points out that a way in which I want to *know*—as a terrible drive with its end enfolded—will, in fact be dramatized in a much larger field in the story. It's like a game, reading this sentence. I see the arm cocked and the point let fly. I get a little blinded by sun and step back. I agitate from foot to foot—then catch it like an ampoule of dye, or poison, or perfume tossed from a speeding sled, safely.

Then, too, there is *this* sentence from *Swann's Way*:

> She was genuinely fond of us; she would have enjoyed the long luxury of weeping for our untimely decease; coming at a moment when she felt "well" and was not in a perspiration, the news that the house was being destroyed by a fire, in which all the rest of us had already perished, a fire which, in a little while, would not leave one stone standing upon another, but from which she herself would still have plenty of time to escape without undue haste, provided that she rose at once from her bed, must often have haunted her dreams, as a prospect which combined with the two

> minor advantages of letting her taste the full savour of her affection for us in long years of mourning, and of causing universal stupefaction in the village when she should sally forth to conduct our obsequies, crushed but courageous, moribund but erect, the paramount and priceless boon of forcing her at the right moment, with no time to be lost, no room for weakening hesitations, to go off and spend the summer at her charming farm of Mirougrain, where there was a waterfall.

Waterfall! The cool, mossy relief after the sentence's journey. The long, bumpy roads, perilous switchbacks and travel-dust clouds–washed away instantly! I read faster and faster, breathless, then–"waterfall," in its solidity and inexplicable arrival. Here is a sentence that withstands me, to which I submit, a sentence that couldn't have known its own end when it started, as I cannot know its end as I begin reading. And I am wholly delighted by the jittery plunge I must take. By the mirror the sentence becomes, in which I see my own surprise.

I love a line cast cleanly out, a shape gently filling the neat spot prepared for it.

And I love a veering, careening ride, the ramble and torque and purifying shock of landing hard.

Freud tells the story of taking a summer walk in the country with a "taciturn friend," a "young but already famous poet." They are ambling along, it's August 1915, the war's on, so imagine the overall heaviness of heart in the slow summer air:

> The poet admired the beauty of the scene around us, but felt no joy in it. He was disturbed by the thought that all this beauty was fated to extinction, that it

> would vanish when winter came, like all human beauty and all the beauty and splendor that men have created or may create. All that he would otherwise have loved and admired seemed to be shorn of its worth by the transience which was its doom.

But Freud disputes "the pessimistic poet's view that the transience of what is beautiful involves any loss in its worth." Then he tries to figure out how mourning works—since that *must* be what the two are experiencing, each in his own way, he believes—a mourning over impending death, life's brevity and fragility. Mourning comes to its own "spontaneous end," he reasons. "Mourning consume[s] itself" he says, and leaves us freshened and ready to attach our love to new objects.

I think Freud must have seen many beautiful nests-with-eggs on that walk to come up with this thought. Is there anything more snugly held, more promising a sign of spring than a surprise cluster of eggs in a nest? I imagine they would have been robins' eggs, blue of the beloved book of my childhood, *Little Bear's Mother,* where first I encountered the color in any meaningful way (as backlit morning playground, then dusky sky, then shadow of the mother over the bear) and was held by it, felt some thirst commence, and drank and drank, and felt, at that stream, the never-enough, never, never, never-enough of pleasure held only briefly still (then gone, but refreshed upon reading—*Again! Please, read it again!*) Freud saw—must have seen—a nest, and constituted therein his response, which was a kind of rivulet of blue between his friend's adamant darks. He must have held his own mourning in his own warm hand—that summer marking the first year of the war—until out came an orange-breasted flame from the blue.

(Often I prefer anger in the face of my own various losses. But too, I have my blue robins' eggs, and looking at them makes me content. In fact, I collected a bowlful over the past year on the walks I took, sometimes three a day, to quell the factions, to run the warring out of my body.)

Recently I was walking to the park and, as I dropped the letter I was carrying into the mailbox, I was stilled by the notion, almost a prediction, that I would find a reindeer, a really tiny one about the size of a lemon. This is the way the image came to me: it "popped in" (maybe fell? down from some nest?). Maybe the weather, a very cool June afternoon, encouraged the weird image's arrival. I attempted to exchange the reindeer for something more seasonal, more discernibly trinkety and likely to surface (clover, bottle cap, penny) but the reindeer was stubborn.

I suppose I might dig around a bit, psychewise—perhaps the reindeer is standing in for something delicate and hidden, meaningful in a way I cannot yet understand.

Along the way there were white tulips so robust they reached to my waist. I saw some kind of evergreen whose uppermost branch shot out like a hooked cane into clear sky. Pink azaleas were dulling to brown and looked more like colonies of coral. And the fertile place the reindeer sprang from (swampy? tundral?) offered up another image: a cleanly flensed frog. Now the two images were overlapping: the frog's icy-blue, skinned legs and the whole and intact tiny reindeer.

Then came the smell of gingerbread, though likely I'm misidentifying some flower's perfume, and while this whole sensation took place in summer, many wintry things kept adding up.

To what, though? To what?

I am of two minds about knowing.

What if I thought about the images this way: simply, that they exist out there, and embedded in shifting forms, the tender and violent enter me, the moment's site for such happenings. No irritable reaching after fact and reason, as Keats would say, just Hello Reindeer. Hello Frog. *Your* absolute smallness. *Your* unexplained end.

These images are meaningful/I have no idea what these images mean. And what do I get if I push the real-but-odd pictures up against the nothing-in-hand?

Maybe a glimpse of the blue flame of an egg.

That old man kneeling in the woods, come upon as I was walking, crouched low at a fallen tree, hands pressed together—was he okay, resting like any pilgrim might, his scant belongings bundled, eyes closed and face tilted up? I looked to be sure he was praying and hadn't just fallen, and to see if I should help. It seemed like a loss he was addressing, for he picked the right props—downed tree, rough cane, small parcel—and added to them only himself, a beseeching presence.

But maybe he *had* fallen. As in "from grace."

Sometimes, against one's will, a oneness of meaning creeps in.

Given the choice between, say, a dozen okay chocolates and one small piece of pure Belgian dark, I'll take the smaller, perfect thing. The brief one-time delicacy. It's always been this way with me. I'll eat it at once, no slow rationing-out, and then I'll live with the fleet abundance and the longing.

But, too, I have these perfect T-shirts, so well fitting, falling just-so—a whole drawerful I took such care in collecting—that I resist wearing them for fear of using them up and then not-having.

I'm drawn to the way rust bleeds out around a razor in the rain.

And I want to pick the razor up so no kid will get hurt.

I want the stain to spread.

And I want no one to run in a mess to the doctor (for I, myself, surprise nail-in-the-foot, once had the awful tetanus shots.)

I want this perfect lost-barn tint contained in the blade's corona every time I cross the street to stay *right here*. (And just for fun?—one of the T-shirts I love and keep safe is the color of that rust, precisely.)

———

Recently, while on a walk, I found a letter in the street, handwritten, to a Mrs. G. from Mary D., a Jehovah's Witness, suggesting another visit and scripture reading (quick, only fifteen minutes, she assures) to ease Mrs. G.'s hurt over the loss of her brother. The letter was beautiful, and ended, "I just wanted you to know that I am still out and will be happy to see you whenever you can make the time—and that's usually what we have to do—make time because it seem time just don't allow. Most Sincerely." I am drawn to the handwriting, a combination of script and print, carefully laid down across plain unlined paper and comfortably sloping, the ease of language, the unselfconscious voice set so directly down on the page, an unmediated mind-to-paper move certain of its task. I wish the letter would go on; I want to hear more of this comforting voice.

But at home, I hide when the Witnesses come. I want to be left alone in my godless world. I want not to be exhorted or cajoled or handed one thing more for my own good. I am fed best by what is left behind. Detritus, loved and held, picked through. (No pure dark Belgian here.) I'd do well as a crow or a vulture, cleaning, paring, finding succulent what has been overlooked and is moldering.

The "good word," okay.

But not to have to receive it fresh and from on high.

———

Today it rained for much of the afternoon. It got dark fast, let go a hard, final downpour, and now the streets are clear and sharp-smelling. The light, these long last days of summer, is low enough to jewel and yellow, blur, and now, if I tilt my head, rainbow all the drops hanging from the phone line. It's that the colors weight the drops, slick them with fire and sea-greens in shifts.

I read for sustenance (more than my own lemon-beaded raindrops on the high-wire can give) Proust on asparagus:

> . . . tinged with ultramarine and rosy pink which ran from their heads, finely stippled in mauve and azure, through a series of imperceptible changes to their white feet, still stained a little by the soil of their garden-bed: a rainbow-loveliness that was not of this world. I felt that these celestial hues indicated the presence of exquisite creatures who had been pleased to assume vegetable form . . .

I walk through this rain thinking, at one time I would point this all out to you in person, hold these drops on the wire against those astral stalks, iridesce the water, roll a pearly drop toward you, fray and sift asparagal light. But now you live in another city and you, in another country, and you (who have not yet even made an appearance here) and I no longer speak of such things.

But I want the shine to live. And before I know it, I am offering, tilting into the light and bringing forth . . . something: *fine beads aloft, an abacus of pearls,* say. I'm sowing some new green, but it's for you, Reader, whom I both know and do not know, who both exist and do not exist, who constitute an elsewhere far, further than I can imagine, years, maybe centuries away.

Whose elsewhere is a balm and a comfort.

"Try Our Delicious Pizza"

There on the postcard was my husband's name, carried along on a soft-looking jet stream, sharing the airspace with other bright-colored mechanical bits—probably called "connexions" or "connectorz" because ad-folks know to mess with the language to better appeal to kids. There in the day's mail was his short, balanced name, spelled out with puzzlelike red/yellow/blue parts sailing along, pulled by a friendly, white, airbrushed current. Someone at the Science Center got their lists mixed up and thought he was a child. (Often, he does look like a curious, alert kid—and he *was* a very sweet boy, as I hear it.) In one of my favorite pictures of him, he's around five, playing a recorder, and wearing a wrinkly, striped shirt. There's sheet music on a stand which the others are following, but I'm sure he's just playing by ear. It's that his name on the card is so bright, tender, alert—if it called him, its timbre would be full of harmonies he'd have known even then, and was probably using to embellish the simple song.

It's hard to retain such joyful playing.

When the card arrived the other day, I felt, more than anything, sadness wing in.

And on the Long Island Railroad last month, it came forth like this: the conductor handed my son a gold Souvenir Ticket and punched it ceremoniously, and my son took the ticket, smiling, knowing he was too old for such things (and he is, he's twelve now and taller than I am) but still, he was perfectly gracious. There it was, in my son's hands: the full understanding that others mean well but won't always see you accurately, and what you give in return for their effort is kindness.

The sadness always surprises, hits sidelong, and then files right in. Perhaps there's a groove constructed by years, a worn place ready to receive it. So when an occasion calls it forth, and after the initial shock of arrival, the sadness knows just where to settle.

I don't go in search of it. Why does it find me?

And I don't entertain it for long when it comes; I don't *want* to be sad, but unbidden, it muscles in—by way of train cars stopped in the station, brimming with coal—B&O and Burlington Northern, the basic, everyday, raw-ingredient lines. I'm not inclined to fix on plaintive train whistles, or rusting small towns built too close to the tracks; I'm not even bent on recollecting Scenes of My Life in the Midwest Age Twenty (sensitive, angsty, ISO edgy landscapes, weedy tracks, soulful connections). Rather, it's the way a coal train *shrinks* as I Amtrak by, en route from Baltimore to New York. As if it's holding just a pinch of black dust, a harmless, loamy soil. I had a miniature train set as a kid (from Germany, I'm sure, the parts were lovely, precise and clockworkish), each car the size of a jelly bean, and I could hook them up in different formations and tap the whole line into motion—gently or they'd topple over. For this I needed a smooth kitchen floor. In her time, I read, the little Czarina also had tiny trains, but with ruby headlights and sapphire windows, for amusement during long afternoons, where, on some inlaid table she, too, I imagined, set journeys going with a tap of her royal hand. (With such trains, one conjures a world by humming softly and staring into the distance, past the cars and deep into the table's pattern—or the floor's green and white linoleum—so there emerges an imagined friend and a story: an exile's in process, the dark's all around and the cold, she gets up from her seat, holding her lap dog and . . . etc.). I thought I could inoculate myself against the sadness, even then, by calling it up, turning it over in a light of my own making. Such quiet one needs to hear the smallest wheels turn! How difficult that is to come by now. And my *real* train-in-the-depot, with its coal fresh-blasted from the ruined mountains of West Virginia, supplier of energy for the capital's grid—well, it returns very quickly to full size, and the wrecked towns and poisoned lakes are hitched to it, and that changes considerably how I feel about festive lights ringing the mall, the White House ablaze and the monuments glowing all night, every night of the year.

And how can a thing that's a deathblow to sadness, itself also be a sadness? Once a perfectly balanced, strong, city poplar was busy greening and swelling its buds, ruffling its leaves, over a strung-out woman in a parking lot. It was a bright weekday afternoon and though she was stuffing baggies and needles deep into a hole in the passenger seat, the light through the trees was so gentle, so ancient it made her look like a peasant planting, bent at the waist, with thick legs and hiked skirt; it was a light that buffed the chipped curb, the crass dumpster, so that the illegality of her task seemed mild and familiar. As I walked past, I thought her form beautiful as she inclined to her paraphernalia. For a moment she was part of the finery of spring. And when I heard a low cooing from the back seat, I actually looked for a contented baby (rush basket, lace-blanket, waking into the lemony frosting of sunlight) but saw instead a man pointing at me. Moving his mouth, hunched like a puffed bird, he jabbed his finger and started to yell, then covered his face and told me to get the hell away. And I wasn't frightened, but saddened by the way the greening, bright leaves got erased, just winched out of sight, and a delivery truck intercepted, and everything overrode the blossoming tree.

It's not that this sadness is more profound, more thoroughgoing than anyone else's, or that it resembles misery or afflicts in the way of a devastation. I'm not arguing its superior potency or measuring the relative force of impact. It's that the sadness induces a particular sensation—say, of watching the vast sea/far horizon from an oceanliner while feeling the circumscribed cabin behind you. And here, it's important to note again—this sadness surprises; it's hosted very specifically in some forms and not others.

So for example, absolutely here: in someone who hasn't much to do on a Saturday and so inscribes the day with errands, not wholly unimportant ones, but certainly not pressing. Who draws the errands out, pushing the cart down each aisle, slowly. Whose browsing is full of drummed-up intent, who thinks about birthdays months in advance and adds gifts-to-buy to her shopping list, to justify staying out longer. When a card alone would suffice, or even a call with apologies for lateness.

The sadness comes not in airport leave-takings, not even when teary, divorced parents pass kids between them, or off to the grandparents, those hostage negotiators. But the arrivals: yes, *then.* Kids running into-the-arms-of: *yes.* The abundant scooping up of a child. Of having been scooped, and, by extension, all the current searches for scooping we undertake, each in our way. The absolute presence the moment of scooping convenes—*I am here/this is perfect*—until, into that breaks the thought: *so brief is our time held aloft, before we are set down again.*

It's not to be found on the interstate itself, 5 a.m., dark and empty except for a few long distance truckers and me, en route to the airport—but up there, in the fully fenced pedestrian bridge *over* the interstate: *that* zinged right in. Why? No one was on it, having ended a nightshift, or heading toward a lousy, minimum wage job. No one stood baleful and peering out, fingers hooked through chain-links, awaiting (habit of inmates and bored kids at recess, abbreviation of full-body longing.) The bridge just seemed tired of having to hold. And the fencing a fragile afterthought: *better cage it just in case*. Just in case what? Then came the scenarios, all desperate, all terrible, abraded by morning's empurpled rise. I drove under, thinking such things. That I must not have been alone in my feeling triggered the sadness.

Apples: the fate of certain ones. The other day, a grocery clerk was unpacking Braeburns from Washington State, gorgeous fruits with shiny yellow and red spin-art skins—and he dropped one. Then picked it up and tossed it in a box at his feet. That's all it took. Especially the *thunk*. To grow, to be picked, to travel so far—to end up discarded in the store, that very last port before purchase. I don't know why this moment of sadness couldn't be fixed by adjusting perspective (toward the bright side: see all the apples that *did* make it safely!) It didn't allow for misperception (maybe that's the wipe-and-restock pile), nor did it support much hopeful conjecture (don't worry lady, those go to the homeless shelter). All the efforts of sun, rain, tilling; all the stacking and storing and driving; all the picking by hand, and after a long day, all the heads on bare mattresses, wrists aching, necks sore, pesticides swallowed, crap wages in pockets—made for a particular density of sadness, just the size of a perfect Braeburn.

Sometimes the sadness sidles up, then reroutes. The atmosphere in the coffee place was nice. I was enjoying some free time while traveling, in the company of a good friend I rarely see. We were just settling in to talk when I looked up and saw on the wall behind him three awful bent spoons, badly fastened with Phillips-head screws to a messed-up piece of wood, meant to be used as a set of hooks—for what? dishtowels?—for $19.99! The handwritten price tag dangling on its little string flipped me out. I tried to consider: "it's someone's attempt, it's the best they can do, maybe this is a rehab project and the store's a supporter providing free space for people getting back on their feet. . . ." The clumsiness was *almost* sad. Sad at the edges and for a split second, but there was no sense at all of craft, no eye for proportion, no care for materials. The longer I looked, the angrier I got. *Twenty bucks*! It took someone two minutes to screw together this mockery of "country" which means, in real life, things sheened from use, rubbed with hands, breath and sweat, not shining with bottled, lacquery crap.

And it's not a deflation, either, this sadness. Here's deflation: I could hardly remember my original creek, the one in my head, after seeing the real thing. Once a singular, clear image in mind, it was retrofitted by the smaller, actual one my friend showed me when I finally visited him after so many years. My creek, the one I imagined for him, was farther from town with a hill beyond it that looked out over the first knowledge of the West I stored up (with much help from Laura Ingalls Wilder.) There was a warmth and a fatness to my creek. It swelled easily in rain, was silvery-quick but not too fast, and you could jump it with a good, running start. As my creek rippled on, a warmth blew through the tall grass and up a small hill. And there, where the light sweetly scented the grass, my friend would sit, and I imagined, his wooden chair was weathered dove-brown and each year sunk more comfortably into itself. But at his creek, there's no hill at all, not even a little swelling rise—just a few bends and he takes his own flimsy lawn chair along. And a beer at the end of the week. This creek offered no chance to gaze past it and refine one's longing. No tall grasses with frogs singing to deepen the oncoming dark. There's no *actual* loss here; I can always compose my original creek. It just takes a little longer now to call up. The sadness I'm talking about is way more forward, resourceful, inventive—ambitious in its identification of a site: *there she is,* the sadness says to itself, *on her stomach on the lawn, idly considering a blade of grass . . . I'll give her a minute, then let it rip: these once-bright greens, browning and dulling, these tender roots drying; a slow, stumbling bee near collapsing; scent of fall, torn bird's nest nearby . . . knowing* her, *that's all it'll take.*

This next is awful—awful-ridiculous, but it has all the identifying characteristics of the sadness under review: that unexpected, swift arrival, and way of empatroning me. At the end of the far aisle at Party City (stay with me here), as I was leaving with my dozen birthday balloons—there was the *Camp Rock Scrap Book and Journal.* God-in-the-details and eternities-in-wildflowers, I swear this sadness was of the wrecking ball variety. Page 1: "Think of a person who has supported you and your dream to be a star. Write a letter to that person." I considered buying the book for further study, but a talismanic fear arose—I cannot have this in my house—and I resisted. So I stayed and read more. I couldn't help it. It went on and on with song-writing hints, inspirational sayings. The sadness was instant. It came in the shape of a soft, chubby girl, sitting alone (empty house, kitchen table), filling it all out, too shy to speak much, with a "pretty face" and "what a voice!" Sadness, that some kid with her stash of dreams would set about this task so dutifully. That she'd follow the steps with diligence and believe there's a *way,* and *Hey, I don't know the way, kid!* I kept thinking, and *Let me tell you a secret: nobody does.* That some adult was so well-meaning, so hoping to be *helpful,* so convinced of the wisdom of "writing it down as the first step in making it happen. . . ."

As if writing it down will make anything happen.

And here comes the woman who, outside Party City, stands in the median and begs at the stop light, holding a sign I can't read *at all.* The words are too many, too small and bunched up.

Yes, I have my glasses on.

As if I don't already know what it says. As if the story changes substantially from sign to sign, from begging spot to begging spot, or offers any real variation on "Homeless Please Help." It's that the sign's impossible to read, and, in that way, useless in making her more real. It's just one more reason things aren't working. There she is, trying, and not knowing why even this is failing.

The unhappy family I once lived next to should have made me sad, but didn't. Because the parents and kids recycled a very small repertoire of arguments, and the tone never changed, nor did the duration or pitch. Because their yelling and meanness became merely annoying as I settled in to make dinner, listening to the radio, with a glass of wine, stilled at the center of my life where light lavished its golds on spilled salt, drops of oil, papery garlic skins, and I was given to see such lavishing before dusk came on fully and it was time to clear the scraps away and call everyone to eat. The yelling so contradicted their rowhouse's pastelly façade and cascading flowerboxes, that whenever I passed it, my eye had to work to reject the bright prettiness, and that wasn't sad, just tiring.

But in *National Geographic*, a series of photos of vast rooms full of jewel-colored birds, tagged and ordered by size and kind, that had crashed into buildings, thrown off course by excessive lights at night: yes. The imposition of light as sadness, and the dark as defenseless—whoever thought darkness would need defending, assumed realm of the scary, haven of self-sufficient night creatures, shy, tender-skinned, but flying and crawling, diving and burrowing, rec- and procreating like crazy in private by way of their special radar.

And briefly now (the illustrations are coming in very fast): hard candies. In a cut-glass bowl. The bowl positioned just-so, near the door. Always a-brim which might look like many guests are expected. I hate hard candies, but I take one, and then a few more, just to keep hope's economy moving.

"Hand Picked With Care" on the blue Chiquita banana sticker. That I hadn't considered *care* was involved, that someone might gently, might delicately pick, that maybe care's *real*, not a value enhancer.

And "Rhinestone Cowboy": the song. (Yes, really.) It came out in 1975, but today on the radio, I heard for the first time the actual words—"and offers coming over the phone" which of course the singing cowboy *wants*, he hit the big time, but which also mean (super-obvious modulation to minor) *a lonely life on the road*. The sudden clarity right there in the supermarket, because back then, in fifth grade, I heard it as "And old-folks going over the foam." Which I thought was a metaphor for *the end of life*, and at the time the suggestion of such an easeful ending, the notion of being carried so frothily away made me very wordlessly sad.

And yet, how strangely good to know this sadness maintains itself over decades.

That it concentrates moments. That it refines them.

How generous sadness is. How capacious.

Mostly, though, how startling—so that just when one's settled into a moment, say on a train to visit one's parents, book out, orange peeled, coffee balanced and cooling, there it is, a swift visitation: coal trains in the station offering the sundered, ruined hearts of mountains. Brief five o'clock light. A frayed cuff on a seatmate. Snowy egrets in a marsh the train passes. Some birds stilled in rushes, others fishing in mud. The mud rainbowed with oil.

One spends the ride containing it all.

Augury

That hanging bird in the maple tree: someone might come and cut it down. Or it might stay and dissolve to bone, blowing through seasons, snared in a mess of fishing line. It must have happened just days ago, the bright body's still heavy and pulls the line tight. If taken down, the absence would mean another's discomfort. And the space where it swings, once open again, a measure of someone's breaking point—*a thing too awful to see.* Which is very close to what I feel, rounding the bend down by the lake, finding the goldfinch invisibly strung. Caught plunging *up*, as goldfinches will, bobbing and looping in jittery arcs when alive. How a very wrong thing inverts the world's laws, stills flight and proposes air can hold weight. How weirdly suggestive is hanging and swaying: *this should be fruit,* the form's ends are tapered, the center's a swell of vesicles, ripening. Wind should make of it *windfall* soon. But the coming upon, the space called *come-upon,* with its soft breeze and footpath, torches the idea of harvest, of gleaning. Detonates "just taking a walk." A bird pinned in air is a measure of wrongness. Walking can't counter it. Redirecting attention towards kids won't erase it. Even if moving quickly past—no progress instills; the sight can't be siphoned from the scene. The bird's presence impinges, like bait.

But yellow gets to be glorious, too. And its brightness not wholly awful. Such a yellow scours sight, fattens it. It is uncorruptedly lemonlike. And the sharp bolt of black on the wing shines like a whip of licorice. At the end of the path and around the bend, here's the coming-upon again. The moment itself doesn't close down. Its brightness is not a slamming door. Yellow's not trying to *make up for* the end.

Time crests there.

The weather patterns.

Fish in the lake. Dogs by the shore with laughing kids.

Why must a last moment be made so visible? And held aloft! Why must it dangle, and shift so softly, and keep on making a finality? From it, light rises. On it light settles. Slippery as tallow. Shushing in breeze.

I think it's good to be in a place where thought can't form the usual way, and a familiar scene—a bird-in-a-tree—gets overturned. Dissembled. Made into a precarity. Looked at one way: cornucopic. Tilted another, it goes sepulchral. How close those can be.

Someone might come and cut the bird down. Or I will, tomorrow.

And after the bird's gone, what would be there, as I come through the trees and around the bend—what, besides shots of memory? An arch of branches over the lake? A green frame around a spot of blue sky, rowboats in a fringe of rushes, the cattails and milkweed about to burst—and past the tangle, just the lake again?

Once, that spot worked like a bower. I liked to walk there and pause at the turn, and enter it, and feel contained. Then, into the bower rained a bird. Dropped a bird. Now swings a bird. Hangs a bird. Yellow shines and yellow ripens. Somewhere are sparrows in a field, seen and *watched over* the story goes, and counted, even as they fall. But come stand in this clearing, late afternoon, the still lake fuzzed with gnats in the shade, the oak's heavy green branch overhead, and lean just so. Center the goldfinch in the frame, squint a little, hold in sight—a planet, flecklet, blot on sun.

A ripe pear, a portent, an airless balloon.

A being whose falling was noted, was seen, whose end was tallied (by the hand of, if you believe.)

An occasion for wondering what it feels like to believe.

There Are Things Awry Here

I found a perimeter, thank God, and I'm walking. I'm making an hour of it, finding a way to get my breathing going hard. These four big lots with big box stores must compass a mile. Measuring helps. I am here (quick check: yes, panting and sweaty) but it feels like nowhere, is so without character that the character I am hardly registers at all. So I'll get to work, in the way I know how:

Here is a farmer entering a black field. He's a proper farmer, bowlegged and leathery, with a serviceable rope looped over his arm. But the farmer comes out of a logo'd truck and the rope links up to a ChemLawn can and off he goes to tend the weeds asserting through the blacktop. He pisses I don't know where during his long day in the sun. His hat's a tattered, red, GO BAMA cap. His tin lunchpail is a bag from Popeye's, just down the road (I mean *highway*).

Here is a rancher coming over a rise, backlit and stiff, sure hands on the reins, eye for the dips that would wreck a fetlock. He's nearly cantering over the brown grass, it's already cropped short, but hey, he's on contract, it's the 15th of the month, so he comes to harrow the edge of the lot. The rancher rides masterfully and the mower goes fast; he turns sharply, leans into the bit, and the beast resists not at all.

Here are the animals branded and waiting, they're tired, they stopped where the grass was fresh and a pond provided. It's dusk coming on, a slight chill picking up that turns them toward home, but they don't raise their heads, catch a scent of dog, of round-up coming. The herd's mixed. "MsBob" is all in with "Luvbun" and "GoTide." "Bubbaboy," "Nully," and "Sphinx" are there, too. The stock are purebred Camaros, Explorers, Elantras, Legends. Docile and ragged. Worn, overfed.

More is wrong.

The flags are frozen. They're fifty feet high but don't move in wind and they carry no sentiment, like "these we hoist high over our small town/farm/ranch to keep alive spirit, memory, fervor. . . ." The flags have names: *Ryan's, Outback, Hooters* (best saloon in town, I'd say, judging by all the horses tied up out front). *IHOP. Waffle House. Wal-Mart* on a far—I'd like to say "hill" but that's out of the question, the hill's been dozed, subdued into "rise."

Here is a field between parking lots—real grass and dirt with bottles tossed in, amber longnecks, flat clears of hard stuff. The word "artifact" comes, but it's bumped out by "garbage," the depths are all wrong, and in a matter of weeks it will all be turned over. Not a field's breaking. Not loamy and clod-filled. More Tyvek and tar. By which things are wrapped, laid in, erected. How easily the new names for "seasons" come forth: *undeveloped, developing, development, developed*. Skirting the site I lose options like "fallow," that yearlong rest wherein land regains strength. I'm losing the language for thoughts about gleaning. "Crop" goes to "cropping" as in Photoshop fixing. (And the term "Photoshopping"—wow, *that* gets confusing.)

Here is a farm woman, her shawl held against wind. It's late February in Tuscaloosa, and the tornadoes that hit farther south last week are still lending their kick. She leans into the gust as she crosses, with bags, the black earth (I'm thinking that black *below* tar), the damp earth (I say *earth* out of habit, I see), but it's very well marked, white lines intersect, and the acre or so she's covered (I'm holding on here, with "acre" as measure) is *field* distance, but it's not a field anymore. She's juggling bags and pinning her name tag, she works at the Cobb, the town's multiplex, and she's late for her shift. On my next turn around the series of lots, she'll be behind glass, with money and tickets. Smoothing her hair. Gulping her Big Gulp. Settling. (*Settler. Settlement.* Sigh.)

A bit farther on, here is a mailbox with its red flag flipped up, in front of the Marriott, my closest neighbor (I'm a Hilton-Tuscaloosa guest for the week.) It's a wooden mailbox on a wooden post, which means "rustic"—and truly, it *is* weather-worn. Around each fire hydrant—the hotels here in parking lot land are each fitted with two stumpy blue ones—grows a thicket of bushes. To hide the hydrant. Though in any small town, hydrants are red and freestanding on actual street corners. This greenery means to convey "tended garden." Which makes the hydrant a reverse sort of flower, one that emits water. Which I guess fits the whole upended scene.

Here are four tall trees in a tangly grove—former trees because now they're dead, though a grove, I know, accommodates all forms of growth and decomposition, all cycles and stages. Long, bare branches and rough, broken ones alternate all the way down. It's the kind of ex-tree that might draw an owl (that's what I'm conjuring, a native barred owl), it's got to be full of grubs just beginning to stir, and it offers a safe, clear view of the land. In the air is the scent of burning *something.* Highway and rubber. Diesel and speed. In fact, it's all over—had I known such a smell as a kid, I'd not notice, or only on days when the wind kicked up. Poor farm wife in her booth, her hair tangled and blown. Gusts helping my rancher into his stable, right up the ramp and the tailgate slams shut. And my farmer—he's holding his rope low and firm while it leaks a bright poison as yellow and brief as a corn snake sunning, then startled, then disappearing back into the ground.

Here in the lot is some corrugated cardboard I thought was an animal's vertebrae (sign of hope, life in burrows!). Here the Brinks truck is outside The Cobb, and the driver is armed, as he's been since the beginning of transfers of loot. Here, with a thought to my love up north, I pluck a dandelion (it escaped the farmer), the gesture complete as it's always been—small, flowery symbol of tender missing. I passed a shard of—it looked like pottery (domestic life/human scale!)—but close up was a shorn chunk of thick plastic.

And before the Committee on Irrevocable Mistakes chose this to do to the land—plant tar, seed commerce—*here* was what?

What was here, that a body moved through it?

Back in my room I can't shake the sensation (despite my dandelion in a plastic cup, curtains wide open, basket of apples to naturalize things): a strangeness, an insistence is hovering. The strangeness makes me say aloud to myself—something *had* to be here, something *had been*.

Something made me make stand-ins, cut-outs, cartoons. It made me possessive, led me to say "my rancher, my farmer, my good farmer's wife"—*mine,* because *I* had to make them. From scratch. Out of *something.* Had to make them *look like*. A past. "The past." I conjured clichés (which come fashioned with roots.) I had to make something, because the land couldn't do it. The land gave nothing up. There was no plan, no narrative here, or tether-back-to. Just boxes to eat in. Big boxes for shopping. One boxy theater with nine movies plexed in. The parking lots gaped. Snipped, sprayed and divided. Unpeopled. Tidied for no one.

Real land is never sad in its vastness, lost in its solitude. Left alone, cycles dress and undress it, chill-and-warm so it peaks, hardens, slides, swells. Real land hosts—voles, foxes, cicadas. Fires, moss, thunder. Rolls or gets steep. Sinks, sops and sprouts. But this land didn't read. It babbled the way useless things babble—fuzzy bees with felt smiles, bejeweled and baubly plaques for occasions, ConGRADulation mugs / frames /figurines. Capped, crusted, contained, so laden with stuff—how can it breathe?

Here, surely, went people with thoughts, in the past—and not as I conjured them: fleet, makeshifty odes, dumb stock-assumptions, citified cartoons, with force of wind, vast stretch of blacktop shaping my story of them very poorly. (Points, maybe, for hale traits I imbued: reticence, dignity, industriousness, skill!) My folks were as flat as those cowboy silhouettes slouched up against mailboxes, but the drive to olden them, tie them back to the earth, give them good pastoral work was real.

I'll start over, since this is America, land of beginnings. Since overnight, *here* didn't clarify at all, I'll start again, very simply, with my simple problem:

Here it's February 2008, and I can't figure out how to get my body to land in a land where the present's not speaking. Where stories won't take, and walking is sliding. I found a cadence to quiet the chatter, a word useful for focus and pacing out steps—*"Refuse,"* which I used as both re-FUSE and REF-use, resistance-meets-garbage, iambic / trochaic, sing-songy, buoyant—but alas, it ordered not much. So today I go searching in earnest. To the library first, then around the corner to Special Collections where I blurt my question to the expert on duty: *Near the site of The Cobb—that whole south side of town* ("mess of emptiness" I'm conveying with pauses)—*by the Big K and Hooters* ("that awful nowhere" suggested by sighs), *before all that, what was* there?

Ah, she says, disappearing in the back, then returning with a stack of yellowing magazines. *Here, try these.*

I find a clear table, spread the magazines out and turn the dry pages.

Once it was February 1942 here. It was British Cadet Class 42E at the Alabama Institute of Aeronautics, a wartime flying school operating in cooperation with the U.S. Army Air Corps. Here First Captain Wheeler wrote in *Fins and Flippers*, the cadets' magazine, a note of gratitude to the American trainers "for interpreting their training system in a manner intelligible to we British Cadets."

Here, in their monthly, the cadets and their officers noted the welcomed, small acts of American civility, and laughed over their own displacement (Moors! Tuscaloosa!) all in the literary conventions of the time—yearbooky, vignettish, clean-cut and well-mannered.

And just for a moment, the ugliness recedes (note of gratitude for the cadets, as they hover around, high in the blue, learning dials and gauges and jostling each other; note of—I can't help it—pleasure, as I read to myself and their lovely accents kick in.) The pour of blacktop, the gray icing of curb, I'm being assured now, *isn't* the earth. That's its burnt crust. That's its sackcloth for unholy times, before the Rapture comes and restores, assumes the earth back to woods, fields, shores where I might ramble and stroll—little myth I can't help invoking, which more commonly goes, in my head, wordlessly: *it'll get better, it'll be righted, cleaned and made pure, it will, how bad could it be, see how perfectly blue the sky is!* (That's stock-Lia talking, brightly, brightly because the ugliness hurts, the wincing is constant; that's the me rucksacked up and ready for hiking, neverminding the dark and gathering clouds, grabbing a poncho, and let's go everyone!)

Here, near the Cobb, is the land where Mac wrote, in *Fins and Flippers,* a little piece called "Our London":

> I remember the sun setting over the last rugged corner of Britain in a blaze of crimson magnificence that we saw when the ship sailed in August. I remember seeing the lights of Toronto start to blink from a small island on Lake Ontario. But best of all—I remember London.
>
> Though I am many thousands of miles away, I see

> her constantly, not as she stands now, bruised and battered, but as she was when I spent my adolescent initiation within her walls; and I am sorry that I was not able to appreciate her then as I do now. For in those days, Regent Street just signified to me the roads that led from Piccadilly Circus to Oxford Street. Charing Cross was just a station that served my purpose in going south. The same applied to Fleet Street, Cheapside and Soho, and a host of other fine places. . . .

Here, my students and I are reading Virginia Woolf, who worked in Mac's London, right through the war, this very same war, on her own piece, "A Sketch of the Past." Almost every entry begins with a mere nod to the war outside her window. Instead, it's her past, her lost houses, land, family—whole eras gone, irreplaceably gone—that demand recounting:

> As we sat down to lunch two days ago . . . John came in, looked white about the gills, his pale eyes paler than usual, and said the French have stopped fighting. Today the dictators dictate their terms to France. Meanwhile, on this very hot morning, with a blue bottle buzzing and a toothless organ grinding and the men calling strawberries in the Square, I sit in my room at 37 Mecklenburgh Square and turn to my father. . . .
>
> Yesterday . . . five German raiders passed so close over Monks House that they brushed the tree at the gate. But being alive today, and having a waste hour on my hands . . . I will go on with this story. . . .

Here, it's London for us, when we gather, my students and I, three hours each night to talk about books, language, art—forms of flight, forms of landing. With my cadets, it's getting less strange. All this sitting and reading together helps.

Here, Ryan, one of my hosts, brought me an umbrella since I came unprepared for sudden storms. Here Group Captain Leonard Thorne notes, as I do, the residents of Tuscaloosa's "wonderful hospitality and friendship."

Nights, here, I am much impressed by my Hilton stack-up of pillows. (I can easily be made to feel rich by an abundance of bedclothes, plumping them while watching bad, late-night TV, letting the excess fall to the floor.) Seems I would have played nicely with R.A.F. Hazlehurst who "still thinks a pillow is a weapon and not a headrest." We'd have blurred the room with soft flying weapons. "Born and bred in Derbyshire. Educated at Winchester. 'Dick' to his buddies . . . ," he's the bareheaded one, no leather helmet-and-goggle set, or dress cap all the others wear, in their *Fins and Flippers* photo.

And here is E.G. Gordon, transferred from the Royal Artillery Anti-Aircraft, born in London, educated at Kingsbury School, Middlesex, whose chief sport is boxing, who "claims to be the shortest man in the R.A.F and so lives in constant dread of six-foot blind dates." Here, layered over the land, are his jitters, which, from his photo, it seems he makes light of: his flight cap is precipitously tilted, one side of his mouth hiked, mischievous, laughing, his tie expertly knotted, his meticulous uniform sharp-pressed and not especially diminutive-looking. Whatever he left behind of himself, whatever I sensed on my walk, subatomic, molecularly present—that which I now know to call E.G.—was right *here*. Was so young. In the photo, he's no more than twenty. If he's alive now, he's older than my father.

Here, I walk into class thinking *Really I have nothing to say to these people, the proper study of writing is reading, is well-managed awe, desire to make a thing, stamina for finishing, adoration of language,* and so on about reverie, solitude, etc. Here, sitting down, I'm going over my secret: *I don't want to be inspiring, I just want to write and they, too, should want that—let's all agree to go home and work hard.*

I walk in, I see people with books, stacks of books I've asked them to read. Besides Woolf, there's James Agee (let's take that out class), who lived with the poorest white sharecroppers of Alabama and whose force of nature, *Let Us Now Praise Famous Men,* was published in 1941, as he might add, *Year of Our Lord,* to dignify the event. (*Event.* I choose my word carefully, friends, for, as Agee writes, 'this is a *book* only by necessity . . .' let's turn to page xi. . . .) Now I'm cooking. I, in my flight suit (black sweater and jeans) look into the faces of my cadets. Everyone's eager. We walk to the runway. We find the ignition.

Here, I am escorted to Dreamland for barbeque, and with Brian, my student, eat my first banana pudding. Here, A.T. Grime, Flight Lieutenant, wrote, "anxious experiments are being carried out by the devoted Mr. Davies, our Dietician, with the object of providing an acceptable Yorkshire Pudding à la Tuscaloosa, to satisfy the palates of our gourmets." "This I think is typical," he continued, "of the efforts made by all at this school . . . to make your pilgrimage a memorable experience." The banana pudding is so sweet, so custardy, full of bananas and cakey white fluff, so heavy and childish, if I'd grown up with it, I'd miss it too, when abroad.

Here, in 2008, the assistant in charge of visitors is a "Fifi." Here, in 1942, the novices en route to becoming pilots first class were "Dodos." Fifis and Dodos. What a menagerie this land raised up.

Here, the novices have their own games: flag-football with elaborate e-mail invitations. "As you can see, we at the UAEDFL—University of Alabama English Department Football League—are incredibly dedicated to our sport; we always give 110% and we play hurt . . . come this Saturday at 11 and feel the RUSH." Here the cadets' training program offered "archery, horseshoes, swimming, tennis, tumbling, softball, volleyball, boxing, relays, calisthenics and, for recreation, golf, checkers (chinese and regular), chess, cards, music, reading, singing, and movies." Posted at

10:59 one night: "After incessant whining on the listserv and the occasional snide (yet sheepish!) remark in the graduate student lounge, the English Department comes up with an unbelievable plan to raise money by playing flag-football. . . . Can this rag-tag band of writers, researchers, instructors and critiquers settle their views on Derrida before it's too late?" And stanza one of an eleven-stanza poem, called "Cadence, Exercise" by cadet J. S. Peck goes:

> Throughout the U.S. Armies wide
> Stand formations side by side
> Their contempt they'll never hide
> for Calisthenics!

Here's Flight Lieutenant Garthwaite, R.A.F. Administrative Officer, in his monthly bulletin "Over to London": "To those at home we send our sincerest hopes for the future. Although not on the field of battle ourselves, we do but gird ourselves for the great and final overthrow of Naziism. Let us hope and trust the coming year will see this war through . . . and . . . not a little by the fruits of our learning over here."

Yes, here they learned their recitations: maximum speeds and service ceilings; flight ranges, fuel capacities, and armaments carried by the Arvo Lancasters, Armstrong Whitleys, and Bristol Beaufighters they'd be flying over the skies at home, soon, soon.

Into this vacancy, *something* asserted. Something strange—that is, real—and insistent was here. The land didn't mean to be torn and tar-covered, wasn't meant to sprout stock farmers, farm women, and ranchers. The land asked to be considered, and seriously. The land wanted to speak—past the bunkers of rolled insulation, past the earth-eating backhoes, and yellow concoction my farmer (okay, *working stiff*, bare hands in the poison, then wiping his nose) force-fed the grass. Here the land must have been green by the runways. Some of the big trees still here must have seen it. It must've been lush once, before hotels started turf wars along Marriott/Hilton lines, and thick vines choked the trees, and the tractors came and the hot blacktop poured, so the SKUs of Big K–hundeds of thousands–might take root and flourish.

I was returned–but not to an Eden, for there were airstrips and the screams of takeoffs, supply roads were laid down for fuel and equipment, the contrails of jets streaked the air, burned, scented, inscribed the quiet so the feel of the whole experience–the desire to serve, the fear of serving–would return whenever humidity, fuel, barbeque combined rightly for the novices.

I was returned, but more in this way: someone dreamed of getting the word, high over Berlin, to top-speed it east toward the Polish border, the Führer, *he's there!*, it's the hamlet of Gierloz, fix your sights son, load, steady, and ––. Someone considered the glory, the fame, posing for photos with requisite wounds. Family pride, shining future. The world's gratitude. Because the *boys* must have thought it, because *I* had the thought, it must have been lingering. The thoughts must have held on, hovering, jittery, wanting some rest. But nothing *cadet* was marked on the land, not poems or pudding, jaunty caps, homesickness. Instead, here were lots, grids, boxes, all manner of automata–doors that opened without human touch allowing the body to float right on in and get down to the business of buying.

Here's where the splintered, close barracks were raised–and then razed, plowed under into a new kind of cloverleaf: blacktopped, clovery only from air.

When the land would not speak and my characters failed, when the land was muffled and my characters stock, this piece was born.

Here is my seed. Here is my search, trail, map of convergences.

Here is the thing I made in place of–*what*, exactly?

What did I find myself wanting? Something simple and telling–say a shop revealing the "character of the people upon whom the town depended for its existence. . . ." Even better (all this from Thomas Hardy), a "class of objects displayed in the shop windows . . . Scythes, reap-hooks, sheep-shears, bell-hooks, spades, mattocks, and hoes at the ironmonger's; bee-hives, butter-firkins, churns, milking stools and pails, hay-rakes, field-flagons, and seed-lips at the cooper's; cart-ropes and plough-harnesses at the saddler's; carts, wheel-barrows and mill-gear at the wheelwright's and machinist's; horse-embrocations at the chemist's; at the glover's and leathercutter's, hedging-gloves, thatchers' knee-caps, ploughmen's leggings, villagers' pattens and clogs. . . ." Oh, boots to lace up against scalding and scraping! Commerce boiled, reconstituted—made rhythmic with breath, heavy with being.

I wanted a footpath, a field-edge—a *sidewalk*. People at ease with neighbors and chatting. A simple plaque at the site of—whatever: *Here the cadets of 42E sat to eat their first grits*. Scrap of wing or propeller on the Hilton's faux mantle. *Fins and Flippers* next to every Gideon's Bible.

What did I find? Some Februaries that matched—one then and one now; some novices each with their good fights and good words, their gratitudes, civilities, and homey soft puddings.

I wanted to know what happened here, on land like this.

Now I know.

People learn to fly through it. And then they go home.

Jump

It's a small thing that holds me.

On the sign that reads *Last Death from Jumping or Diving from Bridge, June 15, 1995,* it's the *or* I can't shake. Why fuss with ambivalence when real mystery abides: here stood intolerable grief or failure. Sheerest abandon, joy in a long summer evening. A dare. Need for adventure/a history of. Why work at precision when, hitched as they are to *Death* in this fragment, both *Jump* and *Dive* convey a misjudging of depth, of current, ignorance of rocks below the dark water, and, with "June" added, an insistent sun peaking the river with camouflage ripples. And isn't it *Death* that I, passerby, secret entertainer of edges and precipices, should instead linger over—approaching, riding, then putting behind me the impulse as I cross the bridge, daily this winter?

Someone thought to be personal about it, not slap up an ordinance "By order of" and "with a $$$ fine." No organization (Bridge Jumpers Anon) claimed the sign; it's not a fraternity service project or probationary do-good feat. That unadorned "Death" is no stat-like "fatality." "Or" is a move to cover the bases, and observed here, now, mid-February, the slightest warmth coming on, barest inflection of sweetness in air, the river still frozen—it opens up all kinds of questions.

Imagine the onset of summer in Iowa, each day in June the light and soft air a surprise, a relief from the long winter's cold. It's been twelve years now since the sign's announcement. The bare facts are holding, but time folds the story back into "the past." None of my friends here remember the death. When I stand on the bridge thinking "twelve years ago now" the form of an actual body in air, in water, is vague and the best I can do to buoy the body is *shirt-puffed-in-wind, corona-of-hair-floating-behind.*

Twelve years ago now. Where'd the story go?

One in which no one moved quickly enough. Because *he* was the athlete. Because she, such a practical joker, would surface any minute, any minute for sure. No one moved off the bridge, tearing a path through the tangle of cattails and blackberry to plunge in and help. Or everyone tried, but she was under too long. Or he stood by himself in the early pink dawn, and the act, intended to purify—the cold water awaken, the silence exalt—was planned as a private moment.

Around the sign, around the inconclusive *or*—because of the *or*, the pause it stirs, the space it opens—fragments and conjectures gather: the last person was drunk. The last person, despondent, tied a brick to her ankle. The last person could swim but not well and didn't account for the rain-swollen currents, for a current at all, it looked so mild, as it does now, even in February. The last person was pushed, wasn't ready and twisted around to protest. The last person hit her head on the railing, unconscious before she entered the water. The last person trusted his body, young as it was and accustomed to pleasure. And below were the snarled, sharp nests of dumped cable. Roots of river plants tough as rope. She cut her arm on a broken bottle and fainted and fell over the edge. He misjudged the span's depths and hit the concrete foundation. She didn't imagine construction debris. She thought the vertigo was over for good a long time ago. He looked up to say he was fine, just fine, but his mouth filled with water and he panicked and choked. She jumped, but midair turned it to swan dive—wanting the grace to set her apart, and to best all the plain summer cannonballers.

I'm not doubting it happened; I believe someone died. It's just that the sign complicates, suggests many competing things at once: by "last death," that there had been previous ones. (But those aren't listed.) And how to be sure if the sign-maker kept up with the project, if "last" means "final" and not "last recorded"? Or if the span of twelve years suggests precautions *were* taken—and they worked, problem solved. You might even assume, if you're inclined to optimism, that the sign, in a crude and grim sort of way, is reassuring: that it's now very safe to jump. Or—given the sign's plain-spokenness, its weird departure from officialese—someone got fed up with the jumping and used the occasion to blunt-force the message, to speak to kids "in their own language" and "to this day" (see how solid that phrase, how it makes time behave and ties up the story) the tone is off-kilter and not to be trusted, since, as kids know, authority keeps its ear to the ground and cooks up new methods of sounding native. And so, ahistorical and inconsistent, chummy in ways that feel fake, the sign frays and unfocuses; offers, then snatches away. Which accounts for the queasiness I feel standing before it.

Without a story, the fragments won't settle.

Possibilities crowd in and distract.

Without the stability of a tale-handed-down, one rushes to make things, rushes the blankness as if it were naked, suddenly stripped—indecent, embarrassed. In need.

Conclusions assert.

Stances take root.

Here's one now, a very unpleasant stance that I'd rather let go—but I'm trying to stay alert, catch the forms of response coming in. So, though I'm cringing, I'll present it in full: there on the banks, in the sun, in June, however enticing, *I'd* have been careful. Judged correctly the depths. Known my strength and its limits. *I'd* never have taken such a stupid risk. Because look, right below, how the eddies gather. *Anyone* could see that means sunken stuff's present. It might have worked as a simple jump (I'm leaning over now, calculating: a feet-splayed or bicycling-around kind of jump, to soften the impact) but not as a dive. No way a dive would've worked—and here comes the stance's fullest expression, I feel it, the coo, the assuring, calm sense of righteousness-and-exemption firming up: *she* must have been drunk. That's the kind of dumb thing you do when you're drunk—just jump, crash through the conventional—childish, careless . . .

. . . as if I've never been careless, lit, held by an ocean, a force late at night erasing my path, rolling it, sealing it up behind me: *just come*. As if I've never been successfully beckoned. As if I'd never beckoned myself, oceanically forceful, convinced by desire and absolved by it, sharply alive and powered by very pure, bright shots of impulse.

Such a sign, in all its uncertainty, opens up another way, too, so I might look again at the riverbank, how green and sweet, and tangled with blackberries. The cattails taut and near bursting (I'm working toward a new attitude here, a mildness I hope to cultivate), the sun releasing the loamy, rich scent of days ripening fast. There's been rain and the river's high and quick, and only a little silty. Breeze lifts my hair, my shirt, reaches around, I'm in summer's good hands and some hasp is removed, latches unclicking, sun unfolding white handkerchiefs on water and other commodious tricks of time, flexing, cajoling *here, enter here*. . . .

And here's the new edge I walk up to, new stance to counter the impatient, first one. I'm trailing it, picking through marshes and dunes. It comes forth in this way, by recalling the lighthouse at Cape May Point, New Jersey. The 275 stairs inside are steep and twisting and once you start climbing, you can't turn back, it's too narrow and there are others behind you. In the heat, the scent of iron lifts, scent of all who have rested cheeks, laid foreheads against the burnished handrail, wishing it into a better one, a familiar one along the boardwalk (there, far below on the beach they can't see, leaning as they are against this one, praising its steadiness, hoping it will go on reliably holding).

Last year, after climbing to the top, I couldn't make myself step through the door and onto the walk. Things break. All the time. Unhinge and unbolt. Hairline-crack. Salt air scours and pocks and gnaws down. Hail full-throttles. Sun dries to dust all it touches.

Why must I consider this, daily?

This summer, though, I stepped right out with my son and walked evenly all the way around. I did not look straight down, but neither did I focus on my safe and near cuticles, wrist, wrist hairs, jacket zipper. I walked out and caught the wind, full in the face. No back to the wall, no inching and praying. Last year, I couldn't bear to see my son out there, but wanting to encourage bravery I said nothing and stayed inside and got busy reading a plaque so he might go freely around holding his father's hand, happy in the fierce wind.

But this year—just fine. I don't know why. Except that I've let go of a lot recently. I think it's made me lighter—which might have meant more easily dizzied, more easily lifted and blown away. But instead I got one of those good, hearty paradoxes, one you can hold and gaze at, like *the-emptying-that-fills*, and feel more solid and certain for.

And here, my first stance, the ungenerous one, mingy and full of judgment gives way. Releases. So I might imagine things differently: that he wanted to fly; she was eager to change, not mince through her days; he followed a spoor hopefully, silently—as I have, spoors of thought, wily ones, supple and leading away, leading to all kinds of precipitous points; she wanted to slip the foundering pace of routine; that which overcast him he was ready to shed, or to drop through and be cleansed of; she meant to restore all that lay festering, and let regret go, into the soft and aureate breeze.

Some stories are so much a part of a place, that the place is singed, stained, impressed with their very particular light. A story gone to lore constructs atmosphere, makes up "the place where X happened" and people will, or decidedly won't say they live near it. Will or pointedly will not tell such a story. Such a story is very much like the biggest tree in the yard, whistling, swaying, dropping its envelopes of light through windows and onto the living room floor—you *own* it but don't think in those terms, until someone says, "That's a nice tree you have," and then it hits: how strange to consider "I own a tree," a presence you live with, beside, under. Are shadowed by. That shadows you in. *Real* lore, I mean. Not like the characters and their spectral antics you hear about on touristy ghost walks in old port cities—stories a guide tells for a fee. Of ghosts, I imagine, who are worn thin (thinner than usual ghost-gauziness) by the same nightly shtick, the guide's delivery paced to group shuffling, mystery dosed out, creepiness tuned to hang in the air: "and to this day, no one knows exactly where X. . . ." Stuck in a story gotten not-at-all-right, night after night, ghosts who would otherwise knock about and rattle some rafters for kicks might think it best to stay quiet.

So how to read a sign like this, bent on recording and telling *something*, but not a story. And even now, if I say "recording" I realize how careless the sign is with facts: if it listed fatalities over the years, with zeros included to account for times when no one went over, then it would be clear: *someone* was watching, the totting up would be real. Ongoing. Believable. And "June 15" would register relief, and be more truly a memorial. But the sign is so sketchy, it feels, instead, like attention dropped off and interest waned. And in that way, the jumper/diver, the subject of one particular moment—a moment en route to being tale-worthy—passed out of mind.

But it hasn't passed out of mind. Not for me. The moment, the story, the last death has been nagging.

It's June now. Four months have gone by since I first stood on the bridge and imagined some stories, tried out some stances. The sign's small, no bigger than a sheet of notebook paper; its simple red letters on white metal, its modesty and starkness read differently early or late in the day: when strolling and I know to anticipate it; when hurrying past and it startles. All this time I've been thinking it over, trying to figure out how to read the thing, trying to locate what's been lost and unsaid.

To that end, my field research might go like this:

As soon as I jumped, I regretted it. I could hardly breathe and kept last-ditch praying: "Please be over, please be over." The freefall was awful; it went on forever, though it must've been only seconds. I felt my brain rise against my skull. I felt my ribs shift, my stomach unmoor, my cheeks go loose. I teared up and couldn't see. I heard nothing but wind and couldn't scream.

Or: As soon as I jumped, I hoped it would last. The freefall was amazing and over too soon. I thought *here comes the water*, then everything went blindingly white in sun, the water met me, and disguised as silver pleats in air, waves of late afternoon held aloft, that steepest, most restorative time of day took me in.

Let me assure you, I did jump (or dive—specification is no longer the issue) but not like this. I didn't go over. That *used* to be me. I used to jump in all kinds of ways, from trees and roofs, into slippery scenes, off edges of the known world thinking *let's see what this brings.* But this isn't me now.

I jumped anew. Really far in. I figured the story itself, if found, would offer some solid occasions for reviewing stances I never imagined. Which is really what's most at stake when standing before a story. So to that end, here:

Students used to leap off the bridge all the time, then swim over to a dilapidated dock outside a boathouse on shore and dry off in the sun. The grounds manager I talked with at the university tells me the kids felt it was a romantic kind of thing to do, a rite of passage. But the dock was a mess, falling apart, and students kept tripping as they hopped from the dock to the bank of the river. One day, years ago, while taking a walk, the president saw this going on and ordered the dock removed. It was never replaced, but the bridge jumping continued. And one summer someone *did* drown. My contact doesn't know who. But in trying to think of how to stop the jumping, someone (also unknown to him) came up with that sign as a deterrent. Now that I reminded him, he said, since it *is* in kind of bad shape, he'd talk to maintenance and see if there's a value in keeping it up there or not. He didn't know why kids kept bending it. Why were they so drawn to it? he wondered aloud.

And here's the story I was most looking for—the one that ought to overcast the bridge, crackle down doom like a slash of lightning over the spot, accompany the hunger of overhead ravens, plait through passersby with the threads of fear, loss, gratitude. At least a small wire of sadness ought to work its way in, or breeze chill bare necks, or scent settle into everyone's sweater to mark the occasion: the story of June 15, 1995, from the *Daily Iowan*, written up the following day goes like this: At 5 p.m. in the afternoon, Jonathan V., nineteen, was hanging out on the bridge with friends. He left his work boots and tie-dyed Doors T-shirt on the bank and went up and jumped. When he didn't surface, a friend leaped in to save him, dragged him toward shore, but lost his grip in the steep drop-off close to the bank. Jon was a roofer and lived with a friend's parents, who treated him like their own son.

He sort of filled in for the one who died huffing butane.

He liked adventure, poetry, art. Was kind to the children. One of the girls in the family spoke for the friends, that crowd on the bridge, laughing and drinking in the late June, long summer afternoon, and gathered again at his funeral (which took place, I checked the weather report, on another perfectly composed day). She said that, to everyone there on the bridge, "it seemed like he could get out of the river if he could get in."

I've made a point of moving through the world with very few markers—no tattoos, latest haircut, religious trinkets/charms/icons (no religion, for starters), no messagey T-shirts, brands, brandings/piercings (except ears). These absences, of course, are signs in themselves, but like a turned field, I believe something more nakedly essential fills in the space, suggests things below, stirring. Suggests empty space isn't empty at all. When the body is blank and distractions are few, gestures reveal: without a tattoo, an arm turned shows exchanges of shade and light, the internal swells of exertion/release. The way muscle tightens to counter resistance is available to the eye. Without a tattoo, one can read tilt of head, set of shoulder, tension rising or falling, and not be stopped at the surface by an ever-present joker's grin/dagger/dragon with tail curling forever under a sleeve. A lack of signage enhances mystery. If the sign on the bridge is a bad tattoo, then even a bad tattoo has its interstices, its fleshy moments of relief, though you have to linger and look harder to see them. Even a bad sketch calls the eye to look, to stay, to ask for the story trying to surface.

The land is seeded with incident, marked imperfectly, but even in imperfect signs, stories go on vibrating.

That little "or" set this in motion.

And the absences the sign offered were exactly where I formed my stances (found my scorn, and the lighthouse which softened me; found myself lacking and that I could be bettered.)

Thought about ghosts. Showed you my research.

Assembled the terribly brief facts of a death.

Most moments of the story called "Last Death from

Jumping or Diving" are unmarked still, not fully gathered, barely asserted. Only lightly sketched in. And, too, the moments I've offered here, moments constituting this piece (my own foray into jumping), also remain ill-marked. Broadly imperfect, still largely unfurled. Without extended thinking on "risk." And that whole part about my letting stuff go, and what that might be, what *else* that might mean—that's not really filled in. I know that.

But you can stand before it nonetheless, whatever is still partial or resistant herein. You can stand before it and read, such a sign (memento mori-like, as in "there is much work to do, Lia, keep at it") as I could come up with, here, Baltimore, MD, June 21, 2007, and I'd not be completely ashamed.

Gray

Here's the cathedral, its gray stone, the gray sky, and all the gray, after-rain mottled streets. And the sky is not a cathedral bell, but *also* gray, gray *alongside*, and the icy puddles are not mirrors of sky, though sky resides in bounded ways there. It is not a cathedral tune, this tone, but the way gray wind and stone cloud together. These grays make up the *right now* I am in, as does the sharp uncertainty of what to do with two suddenly free hours and nowhere to be.

All the likenesses gathering, all the things partaking each of the other, being as one, the many-in-all: *no*. Here, beside my uncertainty (where to go, what to do), gray underwing, stone, ice, median grass—just stay, each unto yourself. As you're inclined: hover or seep. Crack, harrow, or blow.

I can tell you, in my uncertainty, I won't be listening to wind in gray branches, and conjuring far off ocean waves. I won't be revising "here's the gray weight of a cold afternoon" to "an afternoon, cold with the weight of gray noon. . . ." I want no gray, arterial side street contracting with old, fraught scenes, and no, no one's absence reconstituted by cold. No snow-sky hardening its stare. No "grayly they pitched their way forward in cold"—how it must have been in pioneer times, gray woolens, gray blankets and buffalo skins, the dimming gray sky a relief from the glare, though it meant, of course, more snow coming; I do not mean to synchronize their gray anticipation with my gray anticipation.

In this singular moment, I'll have no church bells chased to birdcall. No gravely beautiful sidewalk, ice-cracked, with its palette of grays upriding like little headstones. No minor-key wind-hum. No cloud-spire combo of grays rising up. No parable-like breadth to all this, containing, extending, enlarging by grays.

Just: *now*.

All the gray things like only themselves.

It's February in Baltimore, on Mount Royal Avenue. I've just dropped my son at his Saturday art class. It's almost snowing. Each gray thing in its time, in its place, stands just as it is.

Here's the cathedral.

And here I am, outside, giving thanks.

I'm starting by noting every gray thing.

And by *thanks* I mean *I admit I know not what to do, where to go, with all I've been given.*

Advice

Dear,

Why do some men wear such tight pants, and why are they getting tighter these days?

My Friend,

Men wear tight pants because their legs–thighs, calves, ankles–have been long overlooked. Note the poor ankle, stripped bare by socks rubbing. Today's trends, or being in a band serve up an excuse for tight jeans, black, or dark blue, so men might show off a thigh's curve. But more than this, men who would slip into the body of a woman let their pants suggest this, whisper it, the stiff fabric hauled up over hips, which, too, have gone unremarked, slim hips, slight as a girl's. Such a man minds not at all the mocking of his father, wants no handy loop for the hammer, doesn't care to be handy (for this is about legs and not hands), his jeans so tight his friends laugh and say *nice junk, package, stash, man.* So when he sits, there's a fold, a pressure, slight ache at the crease to remind him.

Walking at dusk, the shadow he is passes over benches and curbs, narrows, resembles that of a woman, and again it's that time, years ago now, when he turned sideways and was called by a girl's name, was mistaken for her and he played along—so well, in fact, that it felt not at all like a mistake. He stages now, for himself, double takes late mornings in gardens, slant against buildings at the end of the day. He returns her, there she is, so he's not so alone, she comes back, the steep dark of another, and *he*, scissory, loose-hinged, at home in the ease and expanse of his body, is *she*. No wind billows his cuffs (no cuffs at all, rolled or bunched, fraying, workerish, *these* jeans are skinny as pencils). What is she whispering, so close to him now as he rests on a stoop, bends his knees, makes a lap, brief ambient space for a dog, for a child. . . ? My Friend, they wear their pants tight so as to feel *she's here again*. To quietly, secretly, call her back in.

Dear,

How can I roll around more in nuance and say the fineness of what comes to me, hovering, wordless, what we know to call *thinking*? So often the edges of thought get sheared, tints hardily brightened, rambles clear-cut. The time I need to meander gets claimed, touched, obligated. Then it's tainted. And I'm left with bald statements and gist.

My Friend,

I have a story I want to tell you.

And here, I almost said, "When I learned to shoot . . ." in order to talk about nuance, that fragile state you describe. It was something about holding the stock tight to my shoulder, the surprise taste of oil when I snuck a lick off the barrel—but in bringing the moment into the light, to you, to our readers, a formalness came. I mean, it took form, found a shape much too quickly. "When I learned to shoot . . ." seemed, for a moment, orderly and right as an introduction. But I've shot a gun only twice. The first time, into a blue sky at clay pigeons and my aim was very badly off, and the second on a farm in Poland at cans on a fence, where I hit every one. That was great, but to say "when I learned to shoot" suggests I've kept up—and I haven't.

I have to reorient now, slow down and figure out how to link up your question (I know you struggled to put it together) with my thought—hardly formed, full of promise—about shooting, the taste of gun oil, scrollwork on the stock I ran my nail over, crescent of dirt I scraped from the barrel, sun in the scope, calm of the scope's much-narrowed world, the space there contained, the order and peace unbidden and also unnerving. I'll have to get back to that scattery inkling, or try to shape it anew, either way, overturn that force driving toward statement, toward fixing a point, the point overtaking and bent on sealing up thought . . . and well, yes, that takes *time*. I see what you mean—about the circling and hovering, and how hard it is to get the world to allow it. How difficult to clear space for a ramble. To love time. To get time to love you.

I'll try again. A different route now.

Leaving Chicago a few weeks ago, I saw from the window of the plane, a wall in Lake Michigan. It was parallel to the shore, I couldn't tell how far out—a knuckle's length from so far up, as I closed one eye and measured. It looked like an Etch A Sketch line, stylus-drawn through a silver emulsion. A boat was motoring from shore toward the wall, leaving behind a white wave that dissolved. It was hard to judge speed, but it seemed the boat wanted to sidle up very close, wanted to fold itself into the concrete. As in the airport just this morning, the woman with the prosthetic leg (leg and hip, judging by stiffness) whose skirt was worn through with three little holes where the contraption rubbed, returned to me that sensation of awkward rotation-and-pivot. In the year I wore a body cast, I, too, rubbed holes in the backs of shirts where I leaned against walls and lockers and cars. Small, precise holes where my cast was rough. Seeing that woman, I knew again (anyone might, this isn't clarivoyance) what it was like to be kept far from the bodies of others. "*Those little holes.*" I said it only to myself. I didn't speak the words aloud (nuance needs space to hover and *roll around* as you note) because how would that sound to her: *I know about the holes. Those are my holes*. So close were the holes all these years! Who knew I'd enter them again, that I'd kept them for just this moment so I might seal up the distance between my body and hers.

My Friend, such moments *do* survive. Give them air. Let them play unsupervised in the field of the body. Keep the tasks of the day aside for as long as you can. Feed silence. Invite time. Resist gist.

Dear,

The other day I wanted to give my body away. Why? I'm not, as they used to say, a "loose woman."

My Friend,

Wasn't it you, who wrote a short time ago saying you felt not at all in possession of your body? But that it wasn't death, either, you meant, nor was it another form of detachment or dissociation. And when you were sitting beside a man whose grand loss was known to all, that worst of all losses, a whole family gone instantly, tragically taken . . . wasn't it then that your own surface slid? And you found no reason to dig into why, or interpret, pathologize, justify. You just wanted to give. I read in the paper the other day (yes, this very same paper that runs this column) a strange, then very right-seeming thing. These people who'd been volunteering at a local soup kitchen for seventeen years said, "We almost don't know why we come here, we've been coming here so long. . . ." They call the hungry men "Sir" and the women "Ma'am." They serve up big portions, set places, clear tables, and scrub out the pots. They are not full of pity. Or no longer are. It's just easy, habitual giving and doing.

Wouldn't you want it be, to him, a relief? Wasn't it that your body, just then, needed not one single thing? Only to give, to offer itself. After much generosity of the daily kind (small things matter, too: take in mail for the neighbors, water plants, listen well), your body meant to extend itself *further*. Into. Another. Be *for* another. This is, after all, an advice column. Who writes and asks who hasn't lost something, or isn't afraid of losses to come, or is presently losing and lacks the will to believe it?

Once I sat next to a man on a train whose back didn't work well—it must have been fused, it stayed rigid as he rose from his seat—and he looked to be in great pain. He held his side with one hand and his head with the other; he rested his head against the train window to redistribute the weight and the pressure, but his breath was still fitful. He stretched a little, as much as he could, then angled stiffly back into his seat where he sighed very deeply. And of that relief, I knew this: it's momentary. All that positioning for a moment of respite.

A dose of respite so the wincing would stop, so the loss would cease, is what *you'd* be, right, for the man you just wrote of? A place to lean into and breathe—your hair, if it's long, or your neck with its oceany warmth, scent of grass because we're all going (*really going*, or wanting to go sooner because of the pain), that bit of relief, so pain in its constancy might be put off, it's edges worn softer—you'd *be* that. You'd get to be part of the moment, the site at which even a brief ease asserted.

Yes, Friend, it's criminal to hold back, stay apart, when one might give and give and give. But we've set this up for the greater good. For the worth of other intactnesses, for the sake of family and order, and country, the body is barred from some forms of giving. For all the body learns to bar, *Amen,* we learn to say.

So your useless, beautiful body behaves. You stay still—as anyone might—in the shivery, mutinous light of loss. Light in gimcracks through fall's granite clouds. Light sliding along the bent ribs of pumpkins. Loss translucing the sugar from maples, the tender backlit leaves aflare. Light rashing us all, slow, fretted and grand. Friend, it's hard to imagine the body in pain when it isn't. Or when you're sweating on a subway in August, hard to conjure the distant and soundless cold mornings of winter.

I believe our best work on earth is in service of likeness. I don't know what to call it—moments of interpenetration? To feel the exchange across borders. You're writing, I think, to say how much you want to work for such a cause. Readers, a challenge: hear past your associations with the word *penetrate*; break it down, past the brutish, go back to its origins: "to place within, to enter within, related to *penitus:* interior, in-most, the in-most recesses." To enter, to be entered is a beautiful thing. Though, yes, how hard to contain complications when bodies are involved. Thanks, Friend, for writing.

—

Dear,

I'm writing again. I'm not finished, though your answer was good. You're right. It is hard. Why is it so especially hard to convince others I'd want nothing in return for the body's work. Enough to be the passage through which alleviation moves. Which feels ancient, and clean, like the form of a simple canoe, mano, plinth. Or a very brief poem, a fragment, a moment so full it needs no expounding—Heraclitus' "the harmony past knowing/sounds more deeply than the known" for example.

My Friend,

Perhaps we should consider the aqueducts of Rome for a moment. These days, in Romavecchia, a suburb just north of the city, runners use the precisely spaced arches to mark distance, dogs piss against them, kids slouch and kiss and smoke under the yellowing stone. In the past, in their time, aqueducts filled the baths, fountains, public drinking spouts of Rome, watered terraces, flushed the entire city's sewer system—ah, to make with the body such a system of response, a tonic, balm, respite! (Heraclitus back at you—"Silence, healing.") My Friend, it's a structural question you're asking. How a thing stands up to time, adapts, changes. Shows itself to be a passage, and useful, anew.

Perhaps we are too fixed in our bodies.

This might help:

Once I saw in a new, slick hotel a very mod bathtub with high sides like a big teacup. Anyone would look fragile in it, unformed and diminished by its size. I imagined at rest there many wet bodies, each as tender as the underside of a wrist, that patch where life could be so easily let. In the low light, it suggested the soft milks of Vermeer, cream in the unspiraling peel of the lemon, the lilac and sulfur hanging in air, gem-bright wine in cut roemers blackening, mossy greens pocking the cut wheels of cheese, puckering apples, freshly killed pheasant rainbowing dark corners—in decline, such brimming; in quietude, torsion.

My Friend, in order to contain the event you're discussing—the tending-after, and after, not-wanting—we would have to be different. Wider and broader. And our language would, too. Need and its overtones—desire, ownership, envy—would not be discordant. We'd carry "aftermath" easily with us, lilacs and sulphur shading the scene, the knowledge of clabbering coming on, the turning and souring under our noses, but not yet, not just yet.

If people are happiest when they're useful, then why can't the body be used for good, or lent out as needed, given over, since we're here for such a very short while? Hard question.

———

Dear,

I know others have questions, practical ones, about love and taxes and families and work. Just one more then. I'm sure it's related. Why is it so hard to believe that, as seen from a plane, clouds really can't hold us? I know, because they look thick and solid, they constitute a way of thinking, perpetuate childish thoughts about heaven. Still, it's hard to imagine they won't soften a fall. Such backlit white curves, such pearled, gray-bright heft . . . until your plane cuts right through, and they resist not at all. They just allow passage.

Dear, why are they so unmoved by our passing?

On Luxury

No one's ready.

To sit here on a wooden bench and not have to think of a gunman shattering the train station's bustle, early light, scent of coffee—*that's* a luxury. I forget sometimes. Like I forget having legs. Which is "just a given," we say. But it *is* given. I'm not arguing by God, Luck, or Science, just that it could be otherwise. And "luxury" is its best measure, that unit, *lux,* of illumination, diurnal, in slices, across a pine floor. Ferocious midday, translucing my boy's ear. Jeweling a wineberry. Breaking in surf, and in outgoing ripples recomposing its silvery veil. Here K-9 patrols are making their rounds, sniffing our bags, moving along. People take mild note and return to their business.

Why aren't we ready to think of our peace?

How amazing I've never planned my escape. Quick, let me think: I'd run down to the tracks, jump off the end and hide under the platform. But I needn't do that. Luxury, to read in the *Baltimore Sun* "murder rate," and not have to see the facts of my life recounted therein. Luxury, to read and not follow the phrase down to a bloody, wet stash of drugs, clothes torn and scattered, the whole torrent of shit, junk, paraphernalia jumping the curb, sluicing a front stoop, *my* stoop, the one I'd climb daily returning from work, couldn't scrub clean, with deep cracks where the necklace they shot him for landed. Luxury, to turn back to roll, coffee, paper. To pair *shooting* with *elsewhere*. To let *elsewhere* be faceless and stoopless, miasmic, panegyric, and broken from mainland. Unmapped. Unsketched. Or sketched very badly and broadly—with stylized "alley" and "pile of garbage," "shattered glass," "prostitutes." I can skim fast, skip the rock of my gaze over the headlines, let it grip nothing, be seized by nothing, just skitter across and sink.

Dear stage. Dear props. Dear *National Geographic*-toned urban blight shots: dusk coming on and through one framing link of the sagging chain fence, a slick, backlit rat. A (wide-scope) child-with-soda toddling close to rat/garbage/needle. Parentless on the glittering asphalt. A deepening red-purple centerfold sky, generously layering rooftops with color, forcing that beauty-in-decay wobble, ruin-threshed, redemptive, as night comes on, lavish-yet-stark, in this, the last photo, so we might turn the page and still breathe.

Yes, luxury (in Latin, also: *a vicious indulgence*) looks like bagged pastries, coffee, briefcases. Neat rolling suitcases (I still think they're marvels), redcaps with trolleys helping the oldsters. Benches—a rich, worn mahogany. Walls—of marble and quarried in Sicily. Wall sconces—bronze, and the whole of the interior lined with creamy Rookwood-of-Cincinnati tiles. I'm balancing on the very edge of the Beaux-Arts here, as the well-intentioned music policy promotes this morning's calming selection, the *New World Symphony*. And here comes the English horn's rise-and-stretch moment, all the tender, new, foresty ferns unfurling, slow rambles in meadows, encounters with moss, swallows, silvery waterfall—everything fresh, alighting, awaking. And here I am, among those in the station, arriving/departing. Here and alive. Alive, and recalling how tense that passage when

I played it myself, the exposed intervals not terribly complicated, but treacherous still. As every English horn player knows, careless phrasing at the modulation, or a tempo too slow (opening quarters, especially) tanks the primordial, tips the whole thing into crassness. Even today, I'm nervous hearing it, having been trained to anticipate ruin and adjust.

Here, now, in the station I'm listening hard. As he is, to the music, in a moment of stillness awaiting his train, this beautiful, scruffy conservatory student, en route to New York with his violin. I see his distraction ("*Not this again, not The New World*") then a softening, as he busies himself (*It's Dvořák at least, not Pachelbel's god-awful Canon*) as he takes out his book, travel cup, iPod. Plugs himself in: Mahler, I bet. He's a serious sort. The board flips to "Departures" and he gathers his stuff. There he goes, toward his train. There he goes with the crowd, finding the gate. He's distracted—his girlfriend, audition, apartment. He's not thinking this lightness, this early-bird ramble could be the very last thing he hears.

There he goes—off to Gate E, with that luxury.

Remembering

How do I remember it? I come to the patch of garden first, in the back. Then the little *mud room* I guess it's called around here, just off the kitchen. I enter the kitchen. To the right there's the living room, a cool, open expanse. Wingback chairs. A fireplace? I'll ask him, was there a fireplace. If we sat in the chairs. If we sat on the floor. Was there a rocker. Was there a mantle. If above an upright piano, hung photos of great-greats in gilt frames with thick wavy glass—his pince-nez, her coiled white hair, and were they really there at all. Back in the kitchen, I remember high counters and that he hopped up and sat there, beside the deep porcelain sink. I was always thirsty. The glasses were tall. There must have been chairs and a past era's table, gold-flecked, silver-rimmed or with space-agey darts. How bright and clean it all was. And he was. His neat hair. The curls cropped and tight—or that was the tension's effect. He had a round laugh but his body was hard, there was nothing excess in gesture or feature. All the lines, pushed against, held.

And now here's the strange gift—sixteen years' distance is about to close up. How easy to say *time-and-space*, to know so little of the science behind it, and know still, to employ it. This long stretch of not having seen, this uninterrupted and very pure distance is a measure of—what? Here, soon, at my door with his curious children, will be one who can tell me something of who I was then—who, like me, could not have said at the time, bound as we were by the present, though we still called it "knowing each other." Once we had only moment-to-moment unfolding days. I was not, then, a still point to reflect on. Over the years, time gave me a form. By now I've long been a contemplation.

I remember he was forthright but kind. I don't think he ever said one hurtful thing. The house he took care of, in his father's family for generations, was a respite. A calm place. It called up the phrase "well-appointed," but all that means, or would have meant to me then, was that one could find pins, twine, glue, sandpaper, tacks—small useful things, notions contained in Sucret tins and Savarin cans under the sink. Lining the mud room. There were chores and they seemed to take up his day. He hauled brush and prepared the garden for winter; there were boots for that purpose, and boots for other purposes. And the house had a place for each thing. That he was apart from my life as a student enlarged my understanding of a day. He wore pullovers with many hidden zippers, each sealing a pocket he made precise use of. Much dark blue, against which he appeared even brighter. The house was warm. The rug was braided. Or a braided rug might be imagined into that space. It's that *my* grandmother's house had these rugs, and my past (with toy cars vrooming over and catching the fibers) now meets up with the space I'm opening again, in *his* house—and I, as the site of, the host of that meeting, step back and watch, eye to wool hillocks and pluckable, heavy black stitches.

There was something that hurt him. That was hurting and he was putting away, or falling more fully into, I didn't know which. He didn't know. There was distance between him and the weltering thing. To either side of the wedge was a violence. I sensed there were ropes, the kind in a seafaring cartoon, with a figure plumb in the middle of coils uncoiling fast. And if, as the anchor kept falling, he *stayed put* as my grandmother would say, he'd have been dragged overboard and down. But he wasn't still. He was working fast away from the rope. I thought the untangling would be a long effort. I thought there were things, meanwhile, he shouldn't put up with. A girl he liked who was careless, not worthy, not at all, in my estimation, since he asked my opinion on such things. She dismissed the chores, oil lamp, canning jars. Footstool's crewel work. Antimacassar. Such was the protectiveness I felt for the house, for the house's old things, and for him. Impossible almost for his body to relax. He moved very quickly across tasks, rooms, yards, thought. The brown garden mended things up. As did the clearing and hauling of brush. All my key scenes are of late fall and winter, variations on and responses to cold. Tea maybe. Maybe hot chocolate. He made something for us. What was it? What, in return, did I make?

There were uncertainties, fretted, impacted. A mother was missing. The alarms were far off, but I heard them ringing. Sometimes they clanged. I remember thinking it would likely be rage, that it couldn't be otherwise, but its name was unknown. I wondered how he lived without her. I couldn't have lived without my mother, not then, my mother who held things for me, past things, and returned them to me when I asked. His mother's missing was a form of damage that kept being done. He called it a decision. That he decided distance was best frightened me. Late afternoons, the blank thing was there, in the quiet, ratcheting, winching.

If I press I can find more: a bike, and a helmet—before laws about helmets, so it hung in a specialized air. There was a patch of yellow at the side of the house. Wasn't there a door—to a basement or a pantry? Was this my first pantry? Did we draw the living-room drapes? No. There were no drapes. And why would we draw them, it was always so gray, or snowing, or verging on snow. It's that memory sidles up to a phrase, and a gesture comes along, too—"draw the drapes." It's that the phrase fit, that the room in its spareness and decency would have put it that way. There were curtains instead. They were sheer and when you brushed by, they noticed. Wasn't he beautiful. Didn't he run ahead with a thought, ask, then fill in an answer before I had one. And wasn't there also a counter-impulse, a quick way of rerouting the statement, refining it, offering it again, like a road widened. The thought bettered. Wasn't he intent on *bettering himself.* (Surely I admired that drive, and as surely turned away from the phrase, thinking it too conventional.) And after retracting the interruption, didn't he slow down, chide himself—and *chide,* that little slicing motion, didn't he pare back his impatience. Wasn't he hard on himself. To that reflex, habituated. Merciless, even. *Excoriating* comes. The core and scrape in that word. And the overtones *cuore* and *striate. Consecrate* too, fit to the very ground of that house where he slept, read, cooked, breathed—as I breathed, in my grandmother's house, deeply, the smell of mornings in winter when the heat kicked on and the wood of the stairs and floors expanded, releasing the scent of years when I wasn't. Didn't I want to quiet him. Soften things. Offer some softness. I keep putting a dog there, then taking it back. *Dog* isn't right, but would have been good, with a bed in his room, its sighs companionable at night.

I never saw the whole house. Parts were closed off to keep the heat in—and that fits. He required very little space and the things he needed were close at hand. He reused and used down to the last. Studying at night, we must have opened the fridge and found not much at all. What did I bring? There must have been beer. I think there was oatmeal—in a tin, if a pantry—but not those single rip-open packets. Too wasteful. Too modern. There were stories I followed, harrowing ones, and much I refrained from asking, which, too, was part of the conversation.

It must have been—knowing nothing of his life, then a lot all at once—that I listened for patterns, and to manage the unstable characters/sequences/motives, and mostly, my own disbelief.

Why say *return? I return* to that place. Why construct, of sensation and time, a circle if all along these memories have been here? And doesn't time also unfold, -buckle, -braid? Have I "stepped back in"? I've tried to say "found it again," of the time, but the time wasn't lost. Can the neither-created-nor-destroyed ether suspend couches with dark wooden scroll-work, framed tatting, the red—and what was so red . . . shirts? thermals? complexions? Did these *wait*? Are they *lent*? After all, he lent the very air a bright tint. And I can add in some fireplace heat, tangerine- and lemon-slice flames to warm the white rind of winter outside. He was training for an event, and then stopped. It wasn't an injury. It was something else. He was moving toward something, and also away. The dimensions are folding. He was smart and precise about mechanical things, compasses, knives, the workings of houses and mowers and weather. About the body and some ways that it worked. Not all ways. I think pleasure surprised him.

He sent, yesterday, now that I'm back in town for a while, a picture of himself and his family, but I haven't looked yet. I want to see, right in front of me, his face as I knew it, compose. I want to be part of the reconstitution, like a puddle stilling again after a truck rumbles by. I want (it'll be any day now) to see, in that very first moment, how years compounded, what dailiness built, how the weather of everyday life grew into countenance and bearing—since one can go about picking up toys, shopping, walking the dog bitterly or tenderly, beset by distraction or filled with gratitude. I want to see which stances took and which slipped away, if there are lines I don't recognize, if there are creases I cannot unknow.

Now I remember—of *course,* as soon as you corrected me, or rather, in your gentler way, suggested—it was *I* who sat up on the high kitchen counters, and *you* who stood near, and that was how we adjusted our heights.

And you'd just returned from a cross-country bike trip: *that's* why the bike and you glowed. My sense of gear, and gears all around was right, though hazy. I had the bright bike against the white house, yellow catching a corner of sight. *Orange*, you said, but it took a few seconds and you had to cast back.

You had only a few classes to finish—and that fits the scene, more reasonably shows why you'd be clearing brush at two in the afternoon or dusk, whenever it was I looked up from my books, out the big window and across the two yards. And you're right, it was only two yards, and there was a gate that only partly unhitched, and I had to slip through holding my breath. The neighbors kept buckets for seedpods, last onions, and hard green tomatoes to fry. The dried stalks crunched underfoot in the cold. The frost made things loud and marked my coming to see you with a form of intention I did not have. Or did not recognize. Or could not admit. I was no good with intentions and outcomes then. Your sensibly ordered, manageable tasks—lined up weekly and weekly checked off—made for me a more solid present. And though you couldn't have known it then, in that way, you dispensed a dose of ease.

———

My father once painted and fired a series of Toby mugs just for fun or to sell, no one in my family recalls—fat English gents, with flushed faces and waistcoats—which only recently I've learned to pronounce *wescotts.* My grandmother kept two of these mugs in a phone nook, in a corner of her dining room where, tethered to the heavy black phone, she'd sit in a straight chair and talk. There was a clipboard for appointments and a pen in a holder, silver-and-black, like a slim torpedo. There was a pair of sharp scissors, a letter-opener and pencils in a green cup with a worn velvet bottom. On higher shelves in the nook, a pewter pitcher, a lace doily, a squat crystal vase.

Most of all I liked the cheeks on the Toby mug gents. They were shiny and round—feminine in their invitation. You were rosy enough to have thought of me then as *dark*. *A dark presence*, you said (I was, likely, brooding), in the co-op one afternoon, which is now a well-lit and spiffy place, full of imported meats and good wine—no more dusty rice bins with big metal scoops, or vegetables clumped with fresh farm dirt. And now, while I'm back in town for a few months, the co-op's across the street from my apartment, and again it's where I shop. You described, just the other day, how we met: you were next to me on the check-out line and almost said nothing because, at the time, you didn't know how to say even the simplest things to women, but you think I invited you somehow to speak—and we found we were neighbors and walked home together. I would have had in my bag dried noodle soup packets, fruit, coffee, and chocolate.

And since you have given it so precisely, sure, I can see the whole scene, but with your blue eyes and not my brown ones, for I don't recall the moment I met you. Not one thing about it comes back. I have no beginning in mind to refer to. I'm back in town, and you're back, or rather, have been here all along, and a present moment again knits up its own feel, so that the idea of an end to *this* time right now—and how soon it will pass!—is nothing I can locate either.

The picture I brought along, of my son and me, cheek to cheek, both of us fake serious-looking and about to laugh—when I showed it to your family at lunch, and said "here's my boy"— that was just part of the story. I was thinking, there's my mother's mouth, my father's brow, but mostly, that's my grandmother's chin. In fact, if I tilt up in a mirror, I can see the set of her jaw, her defiance, mock when we teased, or real when angry or retelling a story in which she persevered or struggled to learn resistance to some form of injustice. I remember, sometimes, being annoyed at the gesture. It came from what I considered (at a sulky thirteen) her narrow repertoire of responses. *Predictable,* I'd think. Which, of course, I relied on. It made for me an idea about what a grown-up should be–consistent, dependable in her actions and responses.

My friend, I must have things of *yours* that you've lost. There must be something I can give you to hold and embellish. Something as useful and strangely orienting as a glance in a mirror.

What is it of yours I've been carrying?

The house is still there, but it's no longer your family's. They sold it. "Right out from under me," you'd say, if you were temperamentally inclined toward bitterness. But you're not. I remember, even then, you met dissolution with an energy for order and repair. And given the way your own losses composed, you checked mine like a pulse. I was a thing to watch for signs of bruise, and to care for. The gauging, the measuring you did: take my memory of that, if it's not already an easy possession. How you'd fuss with my scarf, as you did just today before we all went out for a walk in the bitter cold. You tugged it up higher and, I could see, took note of my insufficient beret.

How just now your eye went fast to the small things awry–more soup to be served, plate of cornbread too close to the edge of the table–how you found the shapes of what needed doing and applied precise and practiced responses. How you attended to your very frail mother-in-law–before she could ask for side table, tea, napkin, and blanket–the solicitousness, I remembered, as natural as doing a thing for yourself. Maybe you can use this, too: that you'd get a little faraway when you fussed, the way a mother with many kids fixes one's zipper while eyeing the wet gloves, messy boots, missing sock of the others she's soon to move toward. Your fussing was easy to accept. I felt the matter-of-factness settle over, and submitted. I remember thinking, too, that maybe the care was compensatory, learned as a child, out of necessity, in defense, or as an enticement.

Today at lunch there were many things to attend. That your mother-in-law was unsteady and needed a napkin looked very much like my scarf askew, my too-heavy groceries, some shingles to nail, some compost to pitch. Tasks unfolded. Time got all wavy and particulate at once. It settled. And didn't it unsettle, too.

It made me rub my eyes and blink.

Let me give you this:

That you looked at your mother-in-law, as you must every day, looked right at her and made her thus real, as anyone is more real when seen and attended; that you kept seeing her back into herself, whoever she was, or was just then becoming.

On Tools

That wood won't work. It was chainsawed roughly by someone who had no eye for logs, and now it can't be split. It's piled by the side of the road in squat disks like kinged checkers, game over. I myself do not cut wood, but review this situation through the eyes of a friend whose pretty impressive woodpile I've seen in a photo he sent. All that wood stacked against the shed—*enough for a winter*, I guessed. *Six cords,* he refined. Later came the word *limb* (as a verb) in conversation. And when I said "ax" and he corrected with "maul" (not really a minor adjustment), choice emerged, the field of cutting enlarged and was no longer the simple stock act of cartoons. *Hatchet* arose. I imagined the restive, neatly hung options lining the shed in the photo. ("Splitting mauls come in four-pound increments: eight, twelve, sixteen—mine is sixteen; between maul and ax, the ax is more versatile: slimmer, lighter, sharper. . . . A maul doesn't have to be very sharp. It doesn't cut, it just pries," he writes, going on about functions and choices, since I asked.)

I started to notice my neighbors' woodpiles, the good, precise stacks and the teetery ones; that people use bins, frames, or canvas haulers; supply varies greatly, from modest stacks for the occasional fire to piles for serious heating; kids mess up the piles substantially, borrowing for clubhouses and skateboard jumps. Here in Baltimore, wood arrives by delivery in fall and sits at the curb until people get around to stacking on weekends.

But about those sharp *quarter rounds* (the novice's love of new words in the mouth): *someone* made them happen. If I stick with this a little longer, I might have a chance, in conversation, to use the phrase *put up wood* (like *put up preserves*) now that I know that's what you call it. And with more time still, I could memorize which wood contains the most BTUs per cord, though I'm finding no helpful acronym for (starting with the highest density) Oak, Ash, Maple, Birch, Poplar, Aspen, Pine.

Recently I had a chance to try it myself. It wasn't complicated and I pretty much managed, but awkwardly. My musculature isn't trained for the task; nothing in my body's habituated that way. If I had a reason to practice splitting (like, say, a fireplace) I might become good. But my arcs through air were all a mess. I couldn't control the ax very well—it kept wobbling out of the orbit I intended. The handle felt too long and unbalanced and I couldn't find the space in air that opened for the blade. I suspect that when you're on, there's a groove the blade remembers, and a fissile core awaiting release that calls the right motion down.

Sometimes I look at people (or read certain books) and think (not unkindly, just with disappointment): *oh, a first-time gesture*—say, at a state fair, at one of those booths where guys try to impress their girlfriends by swinging a mock but heavy sledgehammer that, if landed hard enough on the little pad, raises a lever, mercury-in-a-thermometer style, and hits a red bell at the top. Such gestures are clumsy and haven't yet found even a jerky-effective method. They're all just sloppy force asserted. Every now and then someone comes by and it's clear, you can see: he actually *does* this in the course of a day. The language is there, the movements (both the transitional and the primary) are refined or quirky; either way, a system has been worked out. (Of course the carnie running the game can nail the spot and ring the bell, over and over, though he isn't a big guy; he just knows the sweet place, wherever it is, off-center or flat-on. He swings his rubbery sledgehammer up, lets it hang at the zenith just for a second, then the weight of the head angles into its practiced fall. You can't help but envy the guy. The mild, egressive *huh* of breath, how he seems to both find and create the arc, invisible to others. His ease is seductive and even if he *is* reeky, stuporous, snarly, even if swindle underscores his flattery—*come on, big guy, try for the lady*—I fall a little for him; for the forms of effort naturalized, for the fluencies his body knows just the occasion for.)

A tool can so easily be considered wrong or broken (or rigged, if you're in a bad mood at a fair) until one knows how to use it well (finds a grip, a stance that suits, shifts into a callused spot, performs a three-step predance move, swipes a forehead free of sweat). Or it might be considered too heavy when it's not too heavy at all (you just need the recursive angle, and to waggle into the vector that wants you, sidelong or square-on, etc., depending entirely on the way you're keyed to gravity). You might consider the material you're working on bad, rank, unyielding, anomic, unless you come to know its very particular features, which means you can look like a regular person, walking, say, down 5th Avenue in New York but at the same time recite, if called upon to do so, very solid facts about a tool and its use. People who know such stuff, who possess a sensitivity to tools and to the way jobs like to be done, think of such things as daily, as rote, what you learned as a kid. Just work. See below, from a very long letter on wood from the friend who indulges my interest in woodpiles:

Ponderosa pine, unless it's knotty, splits in clean, straight lines, one swing of the maul, usually. A knot is merely[1] a branch that started when the tree was young, near the center, and grew outward with the tree so that it interrupts the grain of the wood with a cross-directional grain and serves to bind the log together. Some wood has crooked grains—cottonwood, Siberian elm—and is so difficult to split I don't bother to cut it. But if a maul doesn't work, you can go to a wedge,[2] which is a slug of metal about a foot long widening from bottom to top that you drive with a sledgehammer. If you try to use a splitting maul on too-knotty/crooked/grained wood, it never penetrates the fibers far enough to split and you can hit all day and never manage it.[3] Ash is more difficult than pine, but usually straight-grained, so it may resist the first stroke[4] and maybe the second, but then suddenly breaks apart. Since oak is the hardest wood, if you get a piece where the grain runs in spirals,[5] it may take a dozen blows[6] to drive

1. "Merely": I think this means to gently show, say, me, that knots have actual origins, are phenomena of growth and not primarily lyrical owls, dragons, volcanoes hiding in paneling, shifting and motile in those moments before sleep. It also reveals (see #2) the utility value of knowing one's materials.
2. Here's one of those things a person would learn very early on; you don't just set up a log and—bam—split it.
3. "This just isn't working" must have come forth, maybe wordlessly, as when one has been laboring for a very long time, and suddenly a shift in consciousness occurs and clears the way for a new approach.
4. Here he must have noted resistance, paused, raised, swung again, and again, until ash in all its peculiarities was known.
5. "This thing is so freaking hard"—but wordlessly again. Or *Shit, this son of a bitch piece. . . .* Here, once, occurred the moment of stopping, wiping a brow and looking closely at what was causing the trouble, as I might look behind me after tripping on an uneven sidewalk to see what occasioned the stumble.
6. Moment of choosing to stay with, to see when, to see if. . . .

> even a wedge far enough in to separate the stump. (No wonder the oak creaks so. The fibers make these small internal noises, so much quieter than the metal-on-metal hammer blow, so much more expressive.)

If you know your stuff, as my friend does, you get to ponder the dialects of wood; you get to put things together about certain fibers being expressive—and that gets to be offhand, parenthetical. Such thoughts are embellishments, and intimacies. Footnotes show discipleship, the drive to get to the bottom of, the urge to make a path by which one comes to inhabit an idea for herself. Inside those parentheses though, a person who knows stuff stops for a minute, reviews the familiar scene afresh. Finds all that he didn't know he knew. Reflects. Surprises himself. Gets, as is often said about such moments of flight and discovery, *poetic*.

Shit's Beautiful

I don't mean this groovily, as in "wow dude, a lot of goodness out there," or as a reprieve ("it could've been a bad scene, but I woke up feeling fine") or as an exhortation to a sad friend to come on and snap out of it. I mean nothing existential, collective, abstract. I mean, exactly, yours and mine.

I hardly know where to start, in praise.

How about with the functioning system: the grinding, lubricating, and dissolving of food en route to the stomach, set below densities of liver and pancreas; the embellishment of gall bladder (branching like a tethered cloud) and appendix (brief, floating archipelago); the large and small intestines folding, switchbacking, before the sideroom of bladder, all ending in the neat funnel of anus. And the systems of deconstructions, mixing, moving, separating, reducing, so as to build anew; nerves instructing muscles to swallow; storage facilities tucked and filling; enzymes and hormones triggering motion/reestablishing stasis; mucosae protecting; vast surfaces of villi absorbing nutrients; and the recognition systems for starch, sugar, fiber, protein, fat, vitamins, salt, and water, each at their appointed moment, extracting, contributing. The final secreting, compacting, and binding. The folds, slopes, and lobes in consort. The duties prescribed. The sequences ordered, so extrusions of matter are detoxified, synthesized, and rolled out in shapes that mimic the form of the body that made them in wellness—the system, the whole, a marvel of containment and timing. Such matter compelled through a body, such abundantly motile matter passing the retiary, the loci, the conundra of folds and slopes and lobes: an amazement.

I'm going to work on the elegance of shit. Why can't it be so? The compact forms expressive, responsive—as architect Louis Sullivan wrote, "A proper building grows naturally, logically, and poetically out of all its conditions."

Indeed, it's hard to know the wealth you're born into, until it's compromised or taken away. Until the crash comes, the doors close, and gates seal against you, the comings and goings no longer freely, unthinkingly performed.

I'm not the kind of person, who, coming home, can't wait to get into her sweats. I'm not a comfy, earth-mother type (though I do make granola and sometimes bread.) In fact, I'd be happy to disappear now and not have you consider my body at all, but since it's my sole point of reference, let me say without too much embellishment: the system, shut down, is a nightmare of ceasings. If you haven't encountered the language yet, it goes *varices, tears, inflammations, fissures. Ruptures. Bleeding (internal and ex-)*. And for others, it's *pouchings, cramps, kinks, misrouted fluids, excessive trapped air,* even *auto-immunological destruction*.

But in the presense of order and balance, when controls persist and mechanisms of transmission maintain and adjust, divide and release, when the billions and billions of microvilli sway and stir for nutrients, when your own two to three gallons of food and liquid process down to a mere twelve ounces of waste a day—what you're left with is shit. I use the unlovely word for it. Of course, there is no lovely word (though we have in abundance the clinical and cloying), but neither am I shying from this one, all the better to force the issue—resistance—into view. All systems in accord—spit, knock wood—what's left is the body's decision: that's enough, and good-bye to the rest, to all we don't need, that, nevertheless, like the sturdiest bridge, the fiftieth sketch, made a path to the moment where the best has been culled, the excess released, all that lent itself in right measure, let go.

I think I have a well-developed sense of modesty. I wear a bathrobe (embroidered with dragons, okay, but still: *coverage*). I don't wear the low-cut, the too-tight to work. This foray into shit is not about *letting it all hang out* as people used to say. It's that gratitude makes a body earnest in its expression: *"Blessed are you, The Architect . . . who shaped the human being with wisdom, making for us all the openings and vessels of the body. It is revealed and known . . . that if one of these passageways be open when it should be closed, or blocked up when it should be free, one could not stay alive or stand before you. . . ."* So goes the morning blessing in the Jewish prayer book: that we might appreciate health in its presence, not only in its absence.

Consider the weird pleasure of eating durians, natto, fermented shark—things custardy-rotten, ammonia-preserved, softened and richening in their decay. The *haut goût* of game, the *faisande*—that quality, ripeness, character so few foods offer today. The pleasure in a chunk of Pont-l'Évêque, of Pié d'Angloys, of Époisses (fantastic—and banned from all public transport in France). A smear of Taleggio on good, hard bread is a training in complexity. We train for complexity on the way *in*, the scents reinterpreted, overriden, desire teased onto a certain lane, the unlikely thing sought, carefully tended, dosed out on special occasions. About such tastes my son once asked: why in life do we avoid them and in cheese actually want them? About madness I recently read: “Afflicted by, and in communion with, a force both fierce and unseen—a force that both chastened and exalted her.” Chastened/exalted—what a fragile state, the awful and ecstatic a tremor apart, the line between them scoured away by the strange light they share. Such delicacies are specially plated, kept apart from other tastes. Priced for discerning. Rare. Seasonal. Cultivated. Strange.

At low tide on Long Island when I was a kid, I’d find all kinds of things washed up on shore. Before all the plastic there was much more beach glass. I collected mostly the buffed blue and green pieces (so bright and jeweled in water and sun, so disappointing at home, dried, on a shelf) but what I really liked to see were the creatures with soft, pale, brownish-blue or pink, slippery bodies. Little worm-forms nestled in tide-pools, sometimes in shells, half in and half out, poking around with their single leg or siphoning up some colloidal treat and exhaling the sandy rest. Here were the tenderest, most secret bodies—clams of all sorts, and scallops whose half-shells suggested a tiny Venus might be nearby. There were barnacles and mussels still attached to their posts, rocks, planks, tangles of seaweed, vital and bubbling, closing up shop when touched, and then after a long while (how such forms taught a child patience) opening and venturing shyly out again.

Oysters are alive in this twitchy way on their half shell, bedded in ice. I order them rarely, and last time I did, I was very restrained. I ate my small portion of Rappahannocks and drank a very cold, bright cava. I held in my mouth, against my back teeth, each plump, quivery body and pushed it around, not chewing exactly, but not rushing it down. Each oyster pooled in a salty ingle, a sensation that quieted the moment, blurred it with sea spray, compacted the wide, oceanic present into a dense, ferrousy body which I then swallowed, and whose essence I held like a breath, as a fullness until I could surface and prepare for the next fleshy wave, and the next and the next.

Such is the pleasure of entry sustained.

Now consider the opposite, also complex, but a pleasure we aren't accustomed to naming. Not unless you're a high school boy, so I'm told, who with friends compares shits, ranking the best one, where and when it was taken. Thus roughening up the sensation of pleasure, but nevertheless, admitting it. I'm not advocating such coarsening or that the rest of us rescind our discretion; just noting the collective silence, so you can nod in assent, continue not-discussing-it, and yet know-what-I-mean. There are so many forms of unsung release, as when a headache lifts, or a fever, and the easing is so terrifically pronounced, though you're simply returning to stasis, your usual, everyday, pain-free state. But daily, should you choose to acknowledge it, such pleasure is available. For a moment, probably morning. Probably brief. Maybe hard-won. Moved on quickly from, to other more mentionable pleasures and tasks of the day.

And here's something else I've often thought, but not said: even the *arrangement* is beautiful. The bends, twists, and dots. Up against bright porcelain. Magnified by and buoyant in water—the hieroglyphs, ciphers, characters (again this requires sufficient fiber/rest/water in homeostatic doses). A good shit is a portrait of health—a study, a fine reproduction. But I won't borrow from art its frame. I don't care to be thought "transgressive," which feels mostly childish these days or at least small, tight, fussy, constricted. What's transgressive anymore? The measure is skewed and tired—oh lovely and dated Duchampian urinal! And anyway, other things interest me more: humor, loss, urgency, doubt, kindness. Amazement. The transgressive calls forth such basic responses: outrage, acceptance, indifference. Arguments about lack of skill (my three-year-old could take a shit and photograph it! My tax money's going to that crap?) Relativism, the eyes of beholders, and Art being "whatever you want it to be." (Bullshit). The question of art's relevancy. The tic of endless referencing (which I guess assures you you're alive, connected up with history, hot in the marketplace of ideas, or at least on stage for a few shining minutes.)

There's *been* a cross submerged in piss. It's been discussed, discussed, discussed. (I agree we should talk, and thanks for the chance, Andres Serrano.) There have been actual bodies-as-sculptures, receiving the carvings of their artist owners (bloody stick figures in Catherine Opie's ample skin, across her chest, across her back—how'd she do *that*?—and in other self-portraits, the artist trussed, leather-masked, and stuck through with pins). As novelties, I'm drawn to such things. But they register, and then: that's it. The *registering* is most of it. They're over so fast. And not the way an espresso is quick—a lovely, rich shock in a very small dose: which is *enough*. Enough is perfect. It meets the body in its need. You don't go looking for one more blast to top the rest, because after "enough," it's no longer good, just jittery, harsh, and stomach-wrecking. "Enough" is frightening because *what's next?* Because you're done. You sit with your pleasure. At a little table. In a certain light. Until you have to turn back to whatever it was you were doing before the need hit.

I suppose extreme things delay the end, create some space. Mean to distract. Mean to extend. I do like an ache to throb or flare: the deep ache of hard, physical work; the pang of missing someone dear, that, if cultivated and tended, keeps presence fresh. But a person who sticks herself through with pins, relies on Art for an awful lot. And isn't it the very frame she's upending that lends credulity to her work? The transgressive artist seems sort of petulant, like a trust fund street-kid. And the work like flipping the bird from a safe, speeding car. I know, I know they're *expanding the field, widening the scope* of the acceptable. And yes, I suppose I'm doing that, too.

But I intend to think about shit without art in hand.

(Before we go any further, I assure you I don't want to touch or play with it—*that* I don't get, though yeah, okay, as long as everyone's happy and no one gets hurt, sure, fine, etc., but no, that's not what I'm into.) It's the coils, curves, loops against white porcelain, underwater figure eights, rolling ranges, peaks, clefts, and croppings—forms so stripped and simple and pleasing . . . blame the abstract expressionists, colorists, potters, collagists, and printmakers I grew up with, the inadvertant training I had in the pleasures of color and form for its own sake. So four green shapes both were and weren't a family walking, pigs in a line, lakes, continents, clouds, pensive, upright, venturesome, kind. I think photographs of shit *would* be beautiful. The moment just before disposing of, stilled. I'd want them to be some substantial size, say 20" x 20" and a study in blue—the blue of the Virgin Mother's robe, ecstasy, heaven, untouched waters, and the background metallic, industrial, lit. But this is not what I mean to do.

I mean to inscribe the end's particular beauty.

Why is the end not beautiful? It's certainly easier to talk about smells, germs, the unclean, and to keep things neatly in their place. And anyway, why bother to *see,* if a thing's on the way out, about to go down the drain.

We think we're put off by, disgusted by shit.

But really, I think we're afraid of the end.

Like you, I, too, step over shit. In pellet form, though, the shiny heaped up look of it—rabbits' and goats'—is pleasing to me, as pleasing as any pile of glittery baubles, or, a cascade of oily espresso beans. Lucent troves of vescicled berries. Anthracite, tumbled and buffed to a sheen. Hard licorice candy. Frog spawn suspended. Caviar mounded in oystershell spoons.

I step over dog shit and curse when I slip and it sludges up the soles of my shoes.

And bird shit: not so interesting, though good luck, my cousins in Italy say, if it lands on your head (how Old World, the sense that luck and misfortune do a very good job of impersonating each other).

Fake shit cracks me up—the kind you can buy in a novelty store and take home and place on a clean kitchen floor. And the long oo's in "doo-doo." And especially the tight-lipped British "poo" with its perfect, open-ended plosive (which is, I confess, what I heard in my head whenever I read the tiresome Pooh books aloud to my son).

Baby shit—that's kind of harsh on the ear isn't it? Well I, too, softened it with some oo's way back when. It's really not bad, just stuff to clean up; and to be honest, it's kind of sweet when it's your own kid's.

So what *is* disgust?

Once I came upon a squirrel, only its tail and hindquarters visible, as if it were diving into a hole in the ground. How odd, I remember thinking. Squirrels don't burrow so deeply in. Even when burying nuts, they're always looking nervously around. I approached the squirrel, which didn't move. The good, solid word "talon" hung in the air, but when I turned the squirrel over (not with my foot, I wouldn't do that), "talon" wasn't right at all. The cut was too clean. The edges of torso were not folding in as they would have, had they been torn by a beak, if entrails were plucked and some ragged bits left. This was the work of a knife or a cleaver. I know because I've done this to chicken.

The squirrel was posed.

It was a joke.

What the pose meant was *someone's humor*. And if that was funny, then I was scared. I felt as if someone was watching me. And laughing in a way I did not understand. I'm fine not understanding some things; I can't begin to rewire a house. But not getting this produced a shiver in me. It came on fiercely and ran through my body. It was smirky, menacing, taunting.

It made me turn away, ashamed.

Jokes are funny because you feel something coming. You

don't know what exactly, but you feel, hovering, a freshness like a little breeze, before it arrives and airs out the scene. The scene builds and holds you and then lets you drop, and you land relieved, safely caught. But behind the scenes here, too much wrong had gone on. The punch line was ill-gotten. A corruption of means. A deformation constructed to make me look twice, mistrust my sight and the familiar gestures of a body at work.

A good joke throws a window wide, a window you've looked through every morning, but suddenly, everything's bright and firm and nothing like what you'd seen there your whole life. You're angled into a new, strange spot. And you're pleased to be shown your oversight. To see in a way that reveals what you missed. I like that form of not-understanding. That being-in-suspension until. It produces something very like awe, evidence of another mind brewing a thing bigger than me. A good joke is holy in its way. But not this—this enjoyed my helplessness. It felt no limits. It would not play.

It did not love the laws of a body.

It defined "disgust" perfectly.

Memo Re: Beach Glass

Beach glass is increasingly rare these days, given the proliferation of plastic tubs and containers, squeeze bottles of lotion, sunscreen and ketchup, Juicy Juice boxes and pouches, and all manner of silvery, pop-top disposables (whose bright kiddy colors dotting the shoreline are as jarring as a pool of antifreeze in a forest). But I'll begin with an overview. Clear glass is traceable, most often, to hard liquor and white-wine bottles—the latter abundant in the warmer months here in the East, tossed or fallen from our decks and pleasure boats, left behind after our beachy celebrations of sun and solstice. The browns are attributable to brewery trash—Buds, Miller Lights, and the common, local, sometimes historically significant varietals (National Bohemian in Baltimore). Stellas, and other upper-end finds, produce chips the color of spectacular Nordic eyes, original-flavor Sucrets, jade beads, backlit aloe. The blues are most highly prized (see recent *New Yorker* cartoon with happy couple on beach, strolling the shore, hand in hand, him saying, "You are my blue beach glass"). Where might the blue originate? Milk of Magnesia, classy vodka, Vicks VapoRub bottles, all crashed against jetties or reefs on their journey to shore. And, too, there's the chance of finding a pressed blue letter or word, indicating a truly old liniment jar, or a blue iodine bottle's thickened corner worn to a platelet, a sort of halved marble, its center swirl gutted and smoothed.

Chips of beach glass are not usually arranged so discreetly (as you've seen in my charts) but rather displayed in a jumble in decorative bowls or casually scattered in large, nonnative clamshells. In bathrooms. In hallways. Mostly liminal spaces. Though so many pieces are thumbnail-sized, I think of them (silently, to myself) as grains. Since they are en route to being such (with proper churnings in tides and scrapes along ocean floors), I preemptively call them by their most evolved–or is it devolved–name. So these grains proceed to our shores with their various characteristics shining forth, making some more collectable than others. The main characteristics (of the object, of the encounter) that a sea-glass comber considers when collecting are noted below. You might commit them to memory by way of the acronym OPE–as in the archaic, poetic "open," c. 1250, which aptly describes both the attitude and eye required for this endeavor:

1. *Opacity*: Have the motion and pressure of waves, abrasions of sand, hydraulics of tides, the peristaltics of passing through (as some must have) various, fishy digestive tracts, worn away the grain's clarity sufficiently? Thus, unlike a diamond (clarity-ranked and produced at great cost to the environment and with much human suffering), it's the working of the natural and beneficently eroding world that we're after here. It's the roughing up that constitutes value, a scraped, worn, and irregular aspect we prize.

2. *Perimeter*: Each piece of beach glass is a study in the ellipse. The ellipse is a broader, more universal form than a circle, though it appears to yearn toward circularity. It labors under our assumption of the circle as higher form, the spherical as enlightened. In

its attraction to waves and their languors, one sees in beach glass the evections of its passage through seas, bays, inlets, estuaries, a kind of microscale record of the moon's effect on tides—in the way insignificant-seeming frogs best register subtle environmental changes. Imagine, for instance, an old jeweler's loupe, its leather case dusty and dried, its lens scratched—an antique: that is, an object whose state of being is deepened over time by true wear, not a self-conscious made-to-look-worn thing. Where once was a clarity and, one assumes, a circularity, there's a smudgy and surprisingly ovule form, all the better to fit the eye socket, pressed into place and held by orbital and suborbital bones. Such a thing wears in a way that reveals a body's peculiarities, as well as the object's own tendency to bend and adapt.

3. Retention of the original act of *Espial*. I mean, these are *hidden* gifts, and to find them takes an eye trained for certain tones, colors and shapes, amid all the purply siren calls of clamshells, of scallop-shell crimps and fractal flutings, the rough iridescence of oystershells, the mussels' seductive, wet, midnight shine. Razor clams bearded with algae. Seaweed with inflatable, poppable pockets (very engaging, highly distracting). Finding beach glass requires focus—a dimming of range, a bounding of perspective. So much so that the actual moment of finding embeds, and each piece retains a tracery of its original spot: here's the green from under the boardwalk; here's the brown from near the lighthouse which we reached after walking for nearly two hours. Here's a most surprising, clear one, found, though nearly camouflaged, in the dry white sand of the upper dunes amid the pebble-sized, extremely smooth and limpid spheres of Cape May diamonds (bits of granite worn in a uniform, pea-shaped way, powerfully enticing, so opposite from the lovely worn volutes of conch and the undulant crescents of glass under consideration here).

Not too many people are patient enough to throw back the young pieces. I do, because without fully-developed features, without the properties sand and tides produce, beach glass is detritus, junk, trash. Sharp, splintery breakage. Time makes it otherwise. *How much* time? *When* is a piece cured? At what point might the hydrodynamics we're studying turn our specimen from waste to prize? All logical questions, though definitive answers would require tracking devices in the wild, centrifuge trials in labs, carbon-dating electron-micro attention, charts and maps and sampling tools, like those Alaskan ice-borers that dig down through seasons of freeze and melt and pull up long cores of time. So while this point, the moment of transmutation from junk-to-value, certainly could be researched, I've found through experience that the eye is the best judge here. The ability to internalize OPE, to make decisions about a specimen—to stoop-and-gather, or pass it up—conjures a state of being well known to b.g. collectors: a kind of happy blankness, a reverie that counters the anomie of one's daily landlocked existence.

This state might best be defined by isolating its component parts: reverence (such as the ocean and its might demands); primitive postures (largely lost to us today: the bending, gathering, and sifting of bits, the sitting back on haunches in an open-hipped squat); an eye for buried brightness; the pulse-quickening moment of finding; the clever manner in which worn glass at once holds and deflects sun; the essential strangeness of a rare blue chip found tucked amid all forms of biotic matter; a sun as heavy as a hand on the back; the esperance of waves; the sand supporting then shifting into troughs around feet; feet sinking into suckholes as waves crash and recede . . . all while the looking goes intently on.

·

And so in response to the many memos submitted and received over the years from colleagues, on such useful topics as the distinctions between ax and maul, the properties of a variety of woods—i.e., the give or resistance of their fibers upon chopping, their relative BTUs (which I'd forgotten is British Thermal and not Burn Time Units)—and to redress my tardy response to a very fine recent memo (discussions scientist Jeffrey Lockwood's research on the sudden disappearance of the locust swarm) which reads, in part: "It is believed that they bred in certain river valleys in Colorado, and when settlers first arrived and plowed up the land for farming, they killed the eggs that were buried in the soil and so inadvertently wiped out the North American locust"—and including this lovely line, too, so rhythmic, so pleasing read aloud: "These locust swarms had a biomass as great as the bison herds and swept over the great plains in regular migrations"—and other memos of note: a short history of personal gun use; urban vs. rural pigeons; the mysticism of glaciers—I submit, here, this brief. Pulled as it is out of thin air, pulled from the place where that-which-we-didn't-know-we-knew abides. Where so much gathers in a rich miasma until called forth by luck, competition (the aforementioned memos were *very* good), an impulse to sketch, itchiness for form, abundance of love for an object, a drive to give small things their due, or the weight of a personal collection piling up, asserting its presence. I submit this memo whose true subject is both a founding tenet and sustaining goal of the whole operation I'm running here, a subject which bears repeating at times of reorganization, challenging times of uncertainty and instability, lest we forget it: the bright uselessness of joyful endeavors.

Two Experiments & a Coda

Street Experiment

All things go. Snow comes, stays, hardens, then melts. Along one block, dropped and unfound, or unsearched-for, things surface in the thaw. First a penny, and I see the ¢ mark in my head—*who writes that anymore?* I know I wrote a lot of them as a kid, in my play stores making play price tags to stick to the fast-sale things. I think, "Nah, leave the penny," but a few steps later there's a nickel, bright as an open eye on the gray stones, and it earns, in that way, my attention. As coin number two it suggests a series, little coin-crumbs leading up to a Susan B. Anthony prize. The intrigues of finding and finding a way to proceed stir together, form an inclination, steps toward. I go back for them both.

Next comes a tiny feather. I have to go back for that, too, since I passed it, thinking, "Too small to be worth it" and "Not of the series." But it's so fresh, so dry in the slush, just-come-down, unmatted and perfect, from somewhere. The proper name, *pin-feather,* attaches and helps; such precision lends weight. At first I thought, "It's enough just to note this," that noting alone would seal the feather in mind—and it's at just this moment that the experiment starts, the challenges firm, the rules finesse: *everything* found in the space of a block will be picked up and kept, and by way of that decision, a synchronic study, *some* kind of picture will emerge. The likelihood of finding stuff asserts, now that I'm saying "experiment," and the awareness of things-to-be-found is a form itself.

The objects are not talismans; they're not the stuff of magical spells. It's better than that. Things are evidence, and the experiment's a form of alertness to take part in. A turning-toward, where slowly, one possibility shades into the next. Moves bear forth things. Thing begets thing.

Come two stalks of silver milkweed, dried and uprooted, blocks from the riverbank where they grow. And with them, the word "berm," from way back, from nowhere—the place where so strangely, precisely, they've landed. Then a red ribbon. A cork. A lanyard crimped into a stiff, uncomfortable curl. Each rough, hempy end is wound like a noose with black plastic. It's the shape of the thing that complicates—too short for a bracelet, too lumpy to mark a place in a book. The strands tooth into each other, grip tightly, and pinch into hard spines at each side. That I can't tell what the lanyard is for makes me off-kilter. I want to drop it and change the rules, but the experiment's already in motion. Now I have a thing that just *is.* As a series of knots, it seems meditative, like the making smooth of a branch with a knife for the hell of it, practice, alleviation of boredom. A tension twists it, suggesting tendon/scar/pod. Broken finger or tail. Dendrite, those sprouting, frayed ends. The urge to go longer, or to flatten asserts. I may be wrong about *meditative*. There are intentions I cannot know. The story's still going, the end still furled. Discovery's current. Now. Right this minute. I pocket the thing and keep walking.

I find a rubbery, fishbone skeleton. It's a cartoonish form with a triangle head, center spine and three requisite ribs intersecting. It's ready to choke a mean cartoon cat. It fits in my hand. How it came loose from a whole is not clear. If it broke from a key chain, its eye isn't ripped and there's no telling grommet; if it fell from a necklace, there's no silver link. Decoration, fishing gear, useless cereal box prize–I can't tell, but into my pocket it goes.

Stubby pencil with hospital logo, bottle cap, and wet envelope come. Then I see–I'm pretty sure–what will be the last thing. There's a lot of traffic today for some reason. My experiment's running along Linn Street between Market and Jefferson. And here I am, at the corner of Jefferson, my planned destination. A woman sees me eyeing the thing and together we laugh at its clear incongruence, right there in the street and so out of season, at how I'm timing my move because the traffic won't stop, and probably, too, because the sun's finally out and the town's finally warming after a very long winter. All this makes the present moment shine, the white golf tee shine on the wet, slushy cobbles and the tether between us, the woman and me, firm up. I let all the cars pass. I dart out for the tee and she applauds as she crosses.

My path is complete now, here at the corner. But crossing with my pocket of loot, I wonder, right off, what to do if I find more stuff. If something comes, will I stop and reach down? Will extending the rules invite disappointment, the little deflations of staying too long, of trying to rekindle? Will overriding the experiment's frame wreck the objects' odd preciousness, stain the control, dilute the results? These questions are part of the experiment, too. If the experiment's over and the boundaries marked off–does that mean it *ends*? If I am not there, won't things still appear? I consider the notion of unclaimed surprises. Will anyone else's eye wander as mine does, and after the pleasures of finding and taking, wonder *now what* and *what's next?* I cross the street. In my pockets, all the stuff jangles.

I think there's more to it. I'm certain there is.

And though I didn't know it then, the experiment really wasn't over.

Months later, at my last dinner party, I gave away all the stuff I found to ten friends. Before they arrived, I laid the things out on my bed and considered which would best suit each person. The objects found their recipients easily. To Brooke, the split hooves of milkweed spilling their down; weird lanyard to Jeremy; rubber fish to Ryan who also was puzzled by its origin. And one by one, as the things left my hands, I saw it: the experiment was still in motion! When the coins first came, they suggested "series." Then "series" broadened. Parameters firmed: *allow everything, take it all in*. Then, when the time came to leave, things came to mean "remember me."

I have to believe the experiment's end is disguised, even now, as "giving away." I have to believe orange fish/feather/lanyard, the experiment itself is under way still, and that I, at my starting point on the cobbled street was no starting point at all. And that night, my room, emptied of stuff was not at all empty.

Silence Experiment

It takes only a moment to decide.

I let the phone ring and ring and don't answer. And now I'm in it.

My breathing is loud. Drinking coffee is loud. Keeping silent is a thing I'm doing with the whole of my body and I hear things anew. I take deeper breaths, the better to expel more air, because voice isn't here to help me sigh, to shape a thought with sound.

Ice breaks loose and slides roughly down the big skylight, and I startle, but let no brisk exhalation escape. Will someone I absolutely *must* talk to come by? Will I scribble comments on my yellow pad in response, or just stay away? Outside, everything is icing over. A guy scrapes his windshield, no gloves at all. With no words at all, I'm thinking that I'd last not three minutes like that before freezing up. It's windy and the trees are swaying stiffly. My breath catches when they bend lower than it seems they can bear. If they snapped, how loud would it be? And ice pelting the skylight, is it lovely, is it lonely to eat uninterrupted by even the possibility of talk? I have terms to abide, rules to attend as I consider these things.

I'm reading an article in the paper about Roger Staubach and realize I don't know how to pronounce his name. I try an "au" then an "ow" in my head. I clear my throat after a jalapeño and hear my voice, way back there, against bone. Without my voice, something else rests, too. Even listening feels loud, especially the radio report about tin masks made for WWI soldiers whose faces were burned off or ruined in combat. Soldiers had not yet evolved instincts for trench warfare and when they stuck their heads up and looked out they got blasted. Mirrors were banned in hospitals. Suicide was rampant. I conjure the sound of knuckles-on-tin-cheek. The disbelieving rap in air. The writer being interviewed, Caroline Alexander, has done an article about the masks for *Smithsonian Magazine*. Her voice is hard, righteous, assured. When asked about new prosthetic technologies, she refuses the shift in conversation, suggests that perhaps it would be better not to make war in the first place. She lets the words sit. She won't fill the quiet. An extra beat slips past the interviewer, into which the author's corrective tone seeps and stills. I very much admire her ferocity. I very much admire the effect of the silence.

The trees move without talking. Not that they were chatty before, but such a conjecture leads to all kinds of what-ifs. Their bent-to-near-snapping leads and leads. What do you say to a tree in peril? In my way I am answering: tension in the back confirming, ache in the neck abiding. According to the plan, I'm not going to speak all weekend.

And, yes, I *am* bored at times. *Boredom is a state in which hope is secretly being negotiated.* I keep that phrase on my bulletin board. A friend of mine advocates boredom for kids, so they might learn to rely on their inner resources. I think it's good to support certain states-of-being, fragile ones like boredom, in danger of being solved or eliminated. Similarly, I fear for aimlessness, restraint, reverie. On my watch list are the sidelong glance, the middle distance, chatting with strangers, frisson, navigating a body by scent, wandering. Anyway, I don't want to stop the experiment now. Phone calls come that can wait, though there's the call from a friend checking to see if I've lost power since the storm picked up and inviting me to his warm house for dinner—and for the night if I want. I do want. But I don't take the call.

By midnight I feel I can't breathe all that well. So I force myself to sleep. And who knows what I say then.

It's the next day—though the experiment makes a kind of continual present. As does the ice, silvering, encasing. I'm wondering what my first word will be. Should I choose, then, to say LOVE, and set that in motion, to send my love a little telegram by way of my voice, by way of a clear stage from which he might feel it, many states and thousands of miles away? What's the one word I will choose to mark and measure the end of my silence? Should I let it surprise me, that first chancy word by which I buy back my voice? I'm of two minds: draw around the event something like a veil, and stage and perform the word—or let it slip out, like a secret under pressure, tired of holding back a force. It might be something like "excuse me" after bumping someone in the tight aisles of the co-op, or "ah" or "um" as I gather my wits to answer a simple question. Or it will be a barely audible sigh, floating a vowel out, "O," for which, now, still, this morning, my breath alone serves richly. In my head, I try on various collisions with Big Expectations.

The snow blowers are a deep bass distraction, and when they still for a moment, regaining their strength, the thick voices of men at work fill in the silence. The sound rises and falls as the machines are crashed into banks of hard snow and the snow draws up into chutes then explodes in an arc like water from an uncrimped garden hose, hose I liked to stand on and fuss with, quietly hidden around the side of the house as my grandmother watered her extravagant roses. In that way, by silencing the running water, I could make her speak: "What?" and "Huh?" and eventually "Hey, someone's up to something." In the garage, made for hiding, post-trickiness, would be the oil-soaked concrete, scent of lime, fertilizer and grass seed, rakes and hoes with worn wooden handles, beach pails and shovels, fringed canvas umbrella, inflatable seahorse, garden chairs whose scratchy weave marked thighs with red lines.

It's a little past noon when I say it: "Hi." And then "No" in response to my housemate's offer of coffee. Two small, clear words. No rallying gems. No symbols or portents. Up in my room, I repeat the words to myself to feel the effect more privately: "Hi." "No." My nose and throat engage and some empty passages fill up with reverb. The words sound good. Younger and rested. More necessary. Relieved of something and freshened.

Coda

I found pearls. I found a diary. I found a black thong and blue condoms in a purse. I found five T-bones still wrapped and frozen. A set of house keys with an address tag nearly led to my first big crime. I found a packet of private stash pictures. I found a thesis red-slashed on each page. I found a cold jay with its heart still beating. I found a phone, an iPod, a joystick. I found tinned caviar perfectly chilled. I found a wallet stuffed with euros, a man's shoe, a Swiss army knife, plane tickets to Prague. I found an old mercury thermometer unbroken, and it confirmed a balmy thirty-five degrees. My name appeared in the cobbles' damp pockmarks. I plucked a single blade of grass, very fresh, very green, from a crack in the sidewalk: the first blade of the year, and *I* found it—amazing!

What loot! Such a cache and a trove!

Tell me then—are these better finds? Are they somehow more than coin/feather/lanyard? Wilder things confer on me—*what?* Lend the experiment undercurrent, scent of an unnamed district/arrondissement/sector, and make the very stones underfoot remarkable? There's a rhythm, I know, a *drive* to this list. This list, though, doesn't it blast my quieter point about discovery—it's ongoingness, the surprise of that? Doesn't all this excitement override thoughts about beginnings and endings—that they're wobbly and unfixed and slip past their boundaries? Is it *not enough* to know that a street with its stuff, its overlooked prizes, curves and bends, makes its way to my eye, my hands for one very rich season, then passes into the hands of others?

As for experiment #2, the words, the "Hi" and the "No"—how my housemate, a visiting philosopher from Italy, hoped for better ones. She wanted the experiment refined and improved. For me to have undertaken it on a sunny day, to have kept the silence rolling for a week. She wanted to see me negotiate harder. And of course the experiment could have been revved, but what happened that weekend moved by degrees. It was about small adjustments and deepening time. Silence in its most animate form. It included sensations, their span from icy, darkened moments to those blowing and flying, cracking and pelting—time and sensation slipping from worn, gray sky to frayed hose, the gray weave below the green casing revealed, the precise pressure I had to apply, all my weight on one foot making sure the hose crossed the concrete path, so I might properly clamp off the water, stop the brown, threaded o-mouth from gushing—or spraying, if my grandmother was using the nozzle to mist her roses somewhere in the long, hot core of summer.

Surely there is a calibration for all this. Surely such moments are worth noting, small as they are, moving forth and retracing, mildly roving. Surely nothing more amped—stop the noise, kill the hype—need happen to make one certain of existing. Existing precisely, existing acutely—as, say, after a fast when eating commences, the tongue rides slowly the slick curve of a green olive, singular morsel, whose skin resists just a little, then gives, and there comes a burst of briny, sharp pleasure. Then the paring away of brisk scraps from rough pit. Rolling the pit. Holding it, shifting—all those tender and ordered attentions.

Then comes a cool sip.

Ice against teeth. Sweat on the glass.

A breath. Conversation.

Abundance dosed out so as not to confound with its rush of riches.

NOTES

The Lustres: Works quoted include *Paul Celan: Poet, Survivor, Jew* (Jim Felstiner); *Specimen Days* (Walt Whitman); "The Prelude" (William Wordsworth); *A Scrap of Time* (Ida Fink); "A Sketch of the Past" (Virginia Woolf).

"Poetry Is a Satisfying of the Desire for Resemblence" (Theme & Variations): The title quotes from Wallace Stevens' essay "Three Academic Pieces." Other works quoted include: *The Aeneid* (tr. Henry Howard, Earl of Surrey); "Canto 81" (Ezra Pound); "Final Soliloquy of the Interior Paramour" (Wallace Stevens).

Against "Gunmetal": I thank Jim Holmes, master gunsmith, for his guidance on firearms.

Street Scene: "petals on a wet, black bough . . ." is from "In a Station of the Metro" (Ezra Pound).

Being of Two Minds: Quoted material is from "On Transience" (Sigmund Freud), and *Middlemarch* (George Eliot).

There Are Things Awry Here: Quoted material is from *The Mayor of Casterbridge* (Thomas Hardy).

Advice: quoted material is from *Heraclitus: Fragments* (tr. Brooks Haxton).

Shit's Beautiful: The line "Afflicted by, and in communion with, a force both fierce and unseen–a force that both chastened and exalted her" is from the essay "When Madness Is in the Wings" by Michelle Nicole Lee, orginally appearing in *The New York Times.*

LIA PURPURA is the author of six collections of essays, poems, and translations. *On Looking* (essays) was a Finalist for the National Book Critics Circle Award. Her other awards include National Endowment for the Arts and Fulbright Fellowships, three Pushcart Prizes, work in *Best American Essays,* the Associated Writing Programs Award in nonfiction, and The Ohio State University Press and The Beatrice Hawley Award in poetry. Recent work has appeared in *Agni, Field, The Georgia Review, Orion, The New Republic, The New Yorker,* and *The Paris Review.* She is Writer in Residence at Loyola University in Baltimore, Maryland, and teaches in the Rainier Writing Workshop MFA Program.

Sarabande Books thanks you for the purchase of this book; we do hope you enjoy it! Founded in 1994 as an independent, nonprofit, literary press, Sarabande publishes poetry, short fiction, and literary nonfiction—genres increasingly neglected by commercial publishers. We are committed to producing beautiful, lasting editions that honor exceptional writing, and to keeping those books in print. If you're interested in further reading, take a moment to browse our website, www.sarabandebooks.org. There you'll find information about other titles; opportunities to contribute to the Sarabande mission; and an abundance of supporting materials including audio, video, a lively blog, and our Sarabande in Education program.